Kerstin Keller

Organic donor-acceptor thin film systems

Kerstin Keller

Organic donor-acceptor thin film systems

towards optimized growth conditions

Südwestdeutscher Verlag für Hochschulschriften

Impressum/Imprint (nur für Deutschland/ only for Germany)
Bibliografische Information der Deutschen Nationalbibliothek: Die Deutsche Nationalbibliothek verzeichnet diese Publikation in der Deutschen Nationalbibliografie; detaillierte bibliografische Daten sind im Internet über http://dnb.d-nb.de abrufbar.

Alle in diesem Buch genannten Marken und Produktnamen unterliegen warenzeichen-, marken- oder patentrechtlichem Schutz bzw. sind Warenzeichen oder eingetragene Warenzeichen der jeweiligen Inhaber. Die Wiedergabe von Marken, Produktnamen, Gebrauchsnamen, Handelsnamen, Warenbezeichnungen u.s.w. in diesem Werk berechtigt auch ohne besondere Kennzeichnung nicht zu der Annahme, dass solche Namen im Sinne der Warenzeichen- und Markenschutzgesetzgebung als frei zu betrachten wären und daher von jedermann benutzt werden dürften.

Verlag: Südwestdeutscher Verlag für Hochschulschriften Aktiengesellschaft & Co. KG
Dudweiler Landstr. 99, 66123 Saarbrücken, Deutschland
Telefon +49 681 37 20 271-1, Telefax +49 681 37 20 271-0
Email: info@svh-verlag.de
Zugl.: Frankfurt, Universität, Diss., 2009

Herstellung in Deutschland:
Schaltungsdienst Lange o.H.G., Berlin
Books on Demand GmbH, Norderstedt
Reha GmbH, Saarbrücken
Amazon Distribution GmbH, Leipzig
ISBN: 978-3-8381-1368-5

Imprint (only for USA, GB)
Bibliographic information published by the Deutsche Nationalbibliothek: The Deutsche Nationalbibliothek lists this publication in the Deutsche Nationalbibliografie; detailed bibliographic data are available in the Internet at http://dnb.d-nb.de.

Any brand names and product names mentioned in this book are subject to trademark, brand or patent protection and are trademarks or registered trademarks of their respective holders. The use of brand names, product names, common names, trade names, product descriptions etc. even without a particular marking in this works is in no way to be construed to mean that such names may be regarded as unrestricted in respect of trademark and brand protection legislation and could thus be used by anyone.

Publisher: Südwestdeutscher Verlag für Hochschulschriften Aktiengesellschaft & Co. KG
Dudweiler Landstr. 99, 66123 Saarbrücken, Germany
Phone +49 681 37 20 271-1, Fax +49 681 37 20 271-0
Email: info@svh-verlag.de

Printed in the U.S.A.
Printed in the U.K. by (see last page)
ISBN: 978-3-8381-1368-5

Copyright © 2010 by the author and Südwestdeutscher Verlag für Hochschulschriften Aktiengesellschaft & Co. KG and licensors
All rights reserved. Saarbrücken 2010

Zusammenfassung

Diese Arbeit beschäftigt sich mit der Präparation von Dünnschichten organischer Donor-Akzeptor-Systeme.

Seit der Entdeckung molekularer Metalle und leitfähiger Polymere hat sich das Forschungsgebiet der organischen Elektronik entwickelt. Ein Ziel ist hierbei die Herstellung vollständig organischer Schaltkreise, bei der die organischen Moleküle als Leiter, Halbleiter und Isolatoren fungieren. Eine Voraussetzung zur Realisierung dieser Schaltkreise ist die Präparation organischer Dünnschichten und zeigt somit die Notwendigkeit optimierter Aufdampf- und Wachstumsbedingungen.

Organische Materialien bieten eine hohe strukturelle Variabilität und auch die Möglichkeit des Zuschneidens eines Moleküls auf spezielle Anwendungen oder Eigenschaften. Hierzu muss jedoch der Zusammenhang zwischen Struktur und Eigenschaften verstanden werden. Die Präparation elektronischer Bauteile bietet hier neben ihrer Funktionalität auch die Möglichkeit der Untersuchung von Materialeigenschaften.

Eine besondere Klasse innerhalb der organischen Moleküle bilden die so genannten Ladungstransfersalze. Sie bestehen aus einem Donor und einem Akzeptor zwischen welchen ein Ladungstransfer auftritt. Dieser führt zu einer partiellen ionischen Bindung, welche die Ursache für interessante elektronische Eigenschaften ist. Ein in diesem Zusammenhang interessantes Molekül ist der Donor Bisethylenedithiotetrathiafulvalen, welcher abgekürzt auch BEDT-TTF oder ET genannt wird. ET bildet eine Vielzahl von Ladungstransfersalzen, wobei ein großer Teil der bisher bekannten Verbindugen Supraleitung zeigt. Zur Untersuchung wurden ET_2-$Cu(NCS)_2$ als organischer Supraleiter und (ET)(TCNQ) ausgewählt. (ET)(TCNQ) tritt in drei Strukturvarianten auf, wobei eine davon einen Metall-Isolator-Übergang aufweist.

Zunächst wurde im Rahmen dieser Arbeit eine Kammer zur Molekularstrahlepitaxie organischer Materialien aufgebaut und diese in ein bestehendes System zur Präparation metallischer Dünnschichten integriert. Weiterhin wurde das gesamte Kammersystem um eine Schleuse mit integrierter Ausheizeinrichtung, eine Kammer zur Aufdampfung metallischer Kontakte mittels Schattenmaskentechnik und eine Sputterkammer erweitert. Mit diesem Kammersystem ist die komplette in-situ Präparation von Dünnschichtbauteilen, wie z.B. Tunnelkontakten oder Feldeffekttransistoren möglich. Kleinste bisher präparierte Strukturgrößen sind 20 μm.

Zur Sublimation der organischen Substanzen wurden verschiedene Typen von Verdampferzellen realisiert, die sich in Aufbau, Material und Größe unterscheiden. Für diese wurden Untersuchungen zur Temperaturverteilung innerhalb der Zellen durchgeführt, sowie Experimente zu den Aufdampfcharakteristika. Bei Vermessung der Temperaturgradienten zeigte sich der bekannte Effekt, dass die Zellöffnung bzw. der Tiegelrand im Allgemeinen kälter ist als der Tiegelboden, von dem das Quellmaterial verdampft wird. Deshalb wurde bei einigen Materialien folglich eine Rekristallisation des Quellmaterials am Tiegelrand beobachtet. Dieses Problem konnte durch ein verändertes Zelldesign behoben werden.

Zur Untersuchung der Aufdampfcharakteristika verschiedener Zelltypen wurde gleichzeitig ein Simulationsprogramm entwickelt. Dieses bietet die Möglichkeit geometrisch eine Tiegelform vorzugeben und die Aufdampfcharakteristik auf einer Probenfläche zu berechnen. Hierbei werden verschiedene Prozesse berücksichtigt: Verdampfen des Quellmaterials, Migration des Quellmaterials entlang der Tiegelwände und Teilchenkollisionen in der Gasphase. Gleichzeitig ist es möglich einen Temperaturgradienten innerhalb der Zelle einzuführen. Die Prozesse sind voneinander unabhängig und können somit getrennt betrachtet, sowie in verschiedenen relativen Stärken kombiniert werden.

Es wurden hauptsächlich drei unterschiedliche Zellgeometrien untersucht:

- die Effusionszelle, welche sich durch einen zylindrischen Tiegel mit hohem Aspektverhältnis von Länge zu Durchmesser auszeichnet,

- die so genannte Knudsenzelle, welche im Gegensatz zur Effusionszelle zusätzlich eine Deckel mit kleiner Öffnung besitzt,

- sowie die neu entwickelte 2-Stufen-Zelle, die eine Kombination der ersten beiden darstellt.

Der Vorteil einer Effusionszelle ist ihr fokussierendes Verhalten, das um so ausgeprägter ist, je größer das Aspektverhältnis ist. Dieses Verhalten konnte auch in der Simulation gezeigt werden. Im Gegensatz hierzu bietet die Knudsenzelle ein nahezu abgeschlossenes Volumen, das sich beim Verdampfen organischer Materialien experimentell als vorteilhaft gezeigt hat, sowie eine sehr gleichmäßige Bedampfung auf einer weit entfernten Probe. Das ideale Verhalten einer Knudsenzelle ist bekannt und folgt dem Kosinusgesetz.

Um die Vorteile beider Zellen zu kombinieren wurde die 2-Stufen-Zelle entwickelt, die im unteren Teil aus einer Knudsenzelle besteht, die ein Volumen im annähernd thermischen Gleichgewicht darstellt, an die sich im oberen Teil eine Tiegelwand zur Fokussierung des Teilchenstrahles anschließt.

Zunächst wurde das Simulationsprogramm einigen grundlegenden Tests unterzogen. Als erstes wurde überprüft, dass ein simulierter Punktstrahler dem Kosinusgesetz gehorcht. Als nächster Schritt wurde eine Effusionszelle simuliert und das Fokussierungsverhalten untersucht. Anschließend wurde ein Vergleich der drei oben genannten Zelltypen durchgeführt. Hierbei wurde eine Vereinfachung angewendet. Da das Verhalten einer Knudsenzelle bekannt ist, nämlich gemäß des Kosinusgesetzes, wurde die Knudsenzelle nicht vollständig simuliert, sondern die Zellöffnung durch einen Punktstrahler bzw. Flächenstrahler ersetzt. Es zeigte sich das erwartete Verhalten: Bei einer Knudsenzelle ist mit steigendem Probenabstand eine starke Aufweitung des Aufdampfprofils und somit eine gleichmäßige Bedampfung zu beobachten, bei kurzen Abständen ist die Bedampfung im Zentralbereich stark überhöht. Die Effusionszelle führt zu einem recht homogen bedampften Zentralbereich auf der Probe, der sich mit steigendem Probenabstand nur leicht aufweitet. Bei der 2-Stufen-Zelle wurden zwei Simulationen mit verschiedener Länge der zweiten Stufe, also der Tiegelwand, durchgeführt. Hierbei wurde festgestellt, dass die Zelle mit langer Tiegelwand ein sehr ähnliches Verhalten wie die Effusionszelle zeigt. Bei der Simulation der kürzeren Zelle ist bei kurzem Probenabstand die zentrale Überhöhung der Schichtdicke stärker sichtbar, genau wie bei einer Knudsenzelle. Bei größerem Probenabstand zeigt sich jedoch ebenfalls ein eher homogener Zentralbereich. Abschließend kann man sagen, dass die Simulation die theoretischen Vorüberlegungen zum Verhalten der 2-Stufen-Zelle voll bestätigt.

Zur Überprüfung der Simulationsergebnisse, wurden mehrere Aufdampfexperimente mit den drei Zelltypen Effusionszelle, Knudsenzelle und 2-Stufen-Zelle durchgeführt. Bedampft wurde jeweils ein Glasmikroskopträger. Anschließend wurde die Dicke des Deponats ortsaufgelöst vermessen. Hierzu wurde ein einfacher Absorptionsaufbau aus einer Photodiode und einem Laser verwendet. Die bisherigen Ergebnisse zeigen gute Übereinstimmung mit der Simulation.

Knudsen- und 2-Stufen-Zelle unterscheiden sich in einem Punkt entscheidend von der Effusionszelle. Da ein nahezu abgeschlossenes Volumen im thermischen Gleichgewicht vorliegt, ist die Berücksichtigung von Stößen zwingend notwendig. Dies wurde auch durch eine Simulation der Knudsengeometrie ohne Stöße gezeigt. Um eine korrekte Simulation der Knudsenzelle durchführen zu können muss ein Stoßmodell entworfen werden. Innerhalb der Simulation wurden zwei separate Modelle entwickelt und verglichen.

Weiterhin wurde das Simulationsprogramm auf die elektronenstrahlinduzierte Deposition (EBID) angewendet und der optimale Einbauwinkel, sowie der Öffnunswinkel der verwendeten Gasinjektionssysteme untersucht.

Der zweite Teil der Arbeit beschäftigt sich mit der Präparation und Charakterisierung organischer Dünnschichten. Um die Präparation durchzuführen mussten für einige Materialien zunächst Kristalle gezüchtet werden. So waren Cu(NCS)$_2$, ET$_2$Cu(NCS)$_2$ und (ET)(TCNQ) nicht kommerziell verfügbar und wurden selbst hergestellt. Bei Cu(NCS)$_2$ und ET$_2$Cu(NCS)$_2$ waren anschließende Aufdampfversuche nicht erfolgreich. Es konnte jeweils nur eine Zersetzung des Quellmaterials nachgewiesen werden. Für die Präparation von (ET)(TCNQ) Dünnschichten wurden drei verschiedene Methoden angewendet. Zunächst wurden ET und TCNQ koverdampft, dann wurde von (ET)(TCNQ) Kristallen verdampft, welche vorher mit Lösungszüchtung gewonnen wurden. Als letztes wurde ein Koverdampfen der (ET)(TCNQ)-Kristalle mit zusätzlichem TCNQ untersucht.

Zur Koverdampfung von reinem ET und TCNQ wurden zunächst Experimente zu den einzelnen Molekülen durchgeführt. Hierbei zeigte sich, dass die Aufdampfraten beider Moleküle sehr gering sind. Für ET wird eine Zersetzung des Materials vermutet, welche sich durch Schwarzfärbung des Quellmaterials äußerte. Diese Zersetzung wurde auch durch Matrix-assisted laser desorption ionization (MALDI) Experimente untersucht, es konnte jedoch kein eindeutiger Nachweis erbracht werden. Aufgrund der möglichen Zersetzung konnte die Aufdampftemperatur und damit die Aufdampfrate nicht beliebig erhöht werden. Gleichzeitig wurde im Ultrahochvakuum eine Desorption einer schon aufgedampften Schicht vom Substrat beobachtet.

Für das TCNQ wurde eine instabile Aufdampfrate festgestellt. Die Aufdampfexperimente wurden jeweils mit einer Quarzmikrowaage verfolgt, welche die Rate detektierte. Hierbei zeigte sich zwar eine Erhöhung der Rate bei Erhöhung der Zelltemperatur, jedoch wurde ein Absinken der Rate bei mehrfachem Bedampfen mit der gleichen Zelltemperatur beobachtet. Ein Grund hierfür war ein Zusammenklumpen des Quellmaterials im Tiegel, weshalb in folgenden Experimenten eine Mischung des Quellmaterials mit Quarzsand vorgenommen wurde.

Die Proben von ET, TCNQ und der Koverdampfung der beiden waren alle mikro-kristallin. Mit dem optischen Mikroskop ließen sich kleine Kristallite erkennen. Form, Farbe und Größe der Kristallite war von Probe zu Probe sehr unterschiedlich. Untersuchungen mit Röntgendiffraktion erwiesen sich aufgrund der Morphologie als schwierig. Erst nach einer Erhöhung der typischen Messzeiten auf mehrere Stunden konnten Ergebnisse erzielt werden.

Als zweite Route wurde die Verdampfung von (ET)(TCNQ) von Kristallen aus Lösungszüchtung untersucht. Ein Grundgedanke hierbei ist, die Zersetzung des ETs zu verhindern, da es hier in einem anderen Kristallverband vorliegt. Es zeigte sich, dass die Verdampfung des (ET)(TCNQ) bei der gleichen Temperatur wie die Verdampfung des reinen ET auftrat. Dies ist ein erster Hinweis, dass sich das (ET)(TCNQ) beim Verdampfen aufspaltet. Mit dieser Methode konnte (ET)(TCNQ) in der monoklinen Phase präpariert werden, jedoch wurde auch einmal das Wachstum von ET alleine in sehr guter Ordnung beobachtet. Dies deutet ebenfalls auf eine Aufspaltung beim Verdampfen hin. Zum Nachweis wurde Röntgendiffraktion eingesetzt, zusätzlich wurden einige Proben mit energiedispersiver Röntgenspektroskopie (EDX) untersucht. EDX erlaubt eine Aussage über die chemische Zusammensetzung einer Probe. Für die Proben in der monoklinen Phase konnten Schwefel und Stickstoff im Verhältnis 2:1 nachgewiesen werden. Schwefel ist exklusiv in ET enthalten, Stickstoff exklusiv in TCNQ. Somit kann über das Schwefel/Stickstoff-Verhältnis direkt auf ein Verhältnis von ET zu TCNQ innerhalb eines Kristalls geschlossen werden, welches in diesem Fall 1:1 beträgt. Die Herstellung der monoklinen Phase von (ET)(TCNQ) bzw. ET konnte über die Substrattemperatur gesteuert werden. Bei hoher Substrattemperatur wurde ET-Wachstum beobachtet. Dies lässt sich durch ein Entfernen des leichtflüchtigeren TCNQs vom Substrat erklären.

Als dritte Variante wurde das zusätzliche Koverdampfen von TCNQ untersucht. Nachdem ein Aufbrechen des (ET)(TCNQ) beim Verdampfen vermutet wird, ist es möglich, dass ein Defizit des leichtflüchtigeren TCNQ am Ort der Probe auftritt, welches das Wachstum von (ET)(TCNQ) negativ beeinflussen könnte. Die bisherigen Ergebnisse geben jedoch keinen eindeutigen Aufschluss über den Einfluss des zusätzlich angebotenen TCNQ.

Als letzter Teil der Arbeit wurden in-situ Tunnelkontakte präpariert. Dies diente einmal zum Test der Schattenmasken, jedoch zweitens auch zum Test eines Messaufbaus für Tunnelspektroskopie. Die Tunnelspektroskopie ist eine Methode zur direkten Untersuchung der Zustandsdichte von Supraleitern. Es wurde also zunächst eine Struktur unter Verwendung einer supraleitenden Elektrode aus Niob hergestellt. Mit dieser wurde ein Tunnelspektrum bei 2 K aufgenommen. Es konnte die charakteristische Energielücke von Niob beobachtet werden. Dies beweist sowohl die erfolgreiche Präparation des Tunnelkontaktes, als auch die Funktionsfähigkeit des Messaufbaus. Als zweiter Schritt wurde eine Struktur mit organischer Elektrode präpariert. Hierbei hat man den Nachteil der schlechten Leitfähigkeit der Organika. Um diese zu umgehen, wurde die konventionelle Tunnelstruktur modifiziert, indem eine zusätzliche Metallelektrode eingeführt wurde. Der Test dieser neuartigen Struktur wurde mit Kupferphthalocyanin als organischem Material durchgeführt, da dieses robust ist und eine geschlossene Schicht präpariert werden kann. Die Messung dieser Struktur, ebenfalls bei 2 K, diente als Nachweis der Funktionsfähigkeit der neu entwickelten Tunnelkontaktstruktur.

Zusammenfassend wurde ein Mehrkammersystem zur Präparation organischer Dünnschichten und Dünnschichtbauteile konzipiert, aufgebaut und getestet. Es wurden verschiedene Verdampferzellen entwickelt und Studien zu den entsprechenden Aufdampfcharakteristika durchgeführt. Präparativ wurde das Schichtwachstum von (ET)(TCNQ) detailliert untersucht und die monokline Phase konnte hergestellt werden. Erste organische Dünnschichtbauteile auf Basis von Kupferphthalocyanin wurden erfolgreich präpariert und mit einem Tunnelspektroskopieaufbau vermessen. Parallel zur experimentellen Arbeit wurde ein Simulationsprogramm entworfen, welches zur Untersuchung der Aufdampfcharakteristika von Verdampferzellen benutzt wurde. Gleichzeitig wurde jedoch auch die Universalität und Erweiterbarkeit des Programmes anhand einer Studie zu Fragestellungen aus dem Gebiet EBID demonstriert.

Contents

1 Charge Transfer   **13**
    1.1 Basic concepts 13
    1.2 CT in the gas phase/on a substrate 15
    1.3 Physical Properties 17
    1.4 Examples 18
        1.4.1 CT in the gas phase 18
        1.4.2 Molecular metals 18
        1.4.3 Organic superconductors 19

2 Materials and Methods   **23**
    2.1 Materials 23
        2.1.1 ET 23
        2.1.2 $Cu(NCS)_2$ 24
        2.1.3 $ET_2Cu(NCS)_2$ 24
        2.1.4 TCNQ 24
        2.1.5 (ET)(TCNQ) 25
    2.2 Characterization methods 28
        2.2.1 X-ray diffraction 28
        2.2.2 MALDI 29
        2.2.3 EDX 29

3 Molecular beam epitaxy system   **31**
    3.1 MMBE-chamber and sample transfer 31
    3.2 Load-Lock 32
    3.3 OMBD-chamber 33
        3.3.1 Sample manipulator and cooling mechanism 33
        3.3.2 Characterization tools 34
    3.4 Sputtering & contact preparation chamber 34
        3.4.1 Contact preparation chamber 35
        3.4.2 Sputtering chamber 36
    3.5 Small MBE chamber 37
    3.6 Test chamber 38

4 Effusion cells — 39
4.1 Cell types and construction — 39
4.1.1 Ceramic tube cell — 39
4.1.2 Free-filament cell — 41
4.1.3 Graphite cell — 41
4.1.4 Mini cell — 42
4.2 Temperature measurements — 43

5 Evaporation Characteristics — 47
5.1 Introduction — 47
5.2 Theory — 47
5.3 Monte Carlo Program — 50
5.3.1 Introduction — 50
5.3.2 Monte Carlo method — 50
5.3.3 Geometry — 51
5.3.4 Program structure — 51
5.3.5 Basic elements — 51
5.3.6 Program flow — 53
5.3.7 Evaporation — 53
5.3.8 Migration — 55
5.3.9 Collisions — 55
5.3.10 Simulation Output — 56
5.3.11 Simulation Results — 56
5.4 Experiments — 64
5.5 Collisions — 67
5.5.1 Mean free path model — 67
5.5.2 Maxwell Boltzmann model — 70
5.5.3 Conclusion — 71

6 Preparation — 73
6.1 Single crystal growth — 73
6.1.1 $Cu(NCS)_2$ — 73
6.1.2 $ET_2Cu(NCS)_2$ — 73
6.1.3 (ET)(TCNQ) — 75
6.2 Thin films — 76
6.2.1 $Cu(NCS)_2$ — 78
6.2.2 ET — 79
6.2.3 TCNQ — 84
6.2.4 (ET)(TCNQ) — 85
6.2.5 $ET_2Cu(NCS)_2$ — 97

7 Electronic Properties — 99
7.1 R(T) measurements — 99
7.1.1 (ET)(TCNQ) single crystals — 99
7.1.2 (ET)(TCNQ) thin films — 99
7.2 Tunnel Measurements — 101
7.2.1 Measurement principle — 103
7.2.2 Planar tunnel junction — 103
7.2.3 Measurement setup — 104
7.2.4 Device preparation — 104
7.2.5 Results — 106

8 Summary and Outlook — 109

A UHV components — 111
A.1 OMBD chamber — 111
A.2 Stencil masks — 112

B Additional simulations — 113
B.1 Migration and temperature gradient — 113
B.2 Temperature gradients — 113
B.3 Tilted substrate — 114

C Sample Overview — 121

D Sputter parameters — 125
D.1 Aluminum sputtering — 125
D.2 Plasma oxidation — 126

List of Figures

1.1	Energy diagram of charge transfer in organics	14
1.2	Schematic potential energy curve for a CT reaction	15
1.3	CT salt formation of Perylene-TCNQ	18
1.4	Generic phase diagram of κ-(ET)X_2	19
1.5	Generic phase diagram of (ET)X_2 compared to high-T_cs	20
2.1	Molecular structure of BEDT-TTF	24
2.2	Molecular structure of TCNQ	25
2.3	Metal-insulator transition of (ET)(TCNQ), β'-phase	26
2.4	Proposed phase diagrams for (ET)(TCNQ), β'-phase	27
2.5	Crystal structure of (ET)(TCNQ), β''-phase	27
2.6	Bruker D8 diffractometer	28
3.1	Sketch of the five-chamber MBE system	32
3.2	3D model of the main chamber	33
3.3	Sample carrier with mounted stencil mask	35
3.4	Plasma stage - photograph and sketch	36
3.5	Side view and bottom view of the sample carrier	37
3.6	Schematic drawing of the test chamber	38
4.1	Sketch of the simplest form of an effusion cell	39
4.2	Components of the ceramic tube cell	40
4.3	Schematic view of a free-filament cell	41
4.4	Photograph of a graphite cell	42
4.5	T-measurement of a free-filament cell with Al_2O_3 crucible	44
4.6	T-measurement of a free-filament cell with graphite crucible	45
4.7	T-measurement of the ceramic tube cell with Al_2O_3 crucible	46
5.1	Sketch of an open cell and a Knudsen cell	48
5.2	Cross section of a Knudsen cell	49
5.3	Cylinder grid	52
5.4	Program structure	54
5.5	Point source vs. isotropic source	57
5.6	Cross sections of the studied crucible geometries	58

5.7	Simulation of an open cell	59
5.8	3D simulation of the cosine law for a Knudsen cell	60
5.9	Geometric model of a 2-stage cell	61
5.10	Comparison of the different cell types	63
5.11	Knudsen cell, plane distance 6 mm	65
5.12	Open cell, plane distance 6 mm	65
5.13	Short 2-stage cell, plane distance 6 mm	66
5.14	Long 2-stage cell, plane distance 6 mm	66
5.15	Simulation of a Knudsen cell without collisions	68
5.16	Open cell with collisions	69
5.17	Knudsen cell with collisions	70
5.18	Knudsen cell with collisions - 2	72
6.1	Powder spectrum of $Cu(NCS)_2$	74
6.2	Powder spectrum of $ET_2Cu(NCS)_2$	75
6.3	(ET)(TCNQ) crystal, grown in DCE	76
6.4	Powder spectra of (ET)(TCNQ)	77
6.5	X-ray diffraction of $Cu(NCS)_2$	78
6.6	MALDI spectrum of ET as purchased	82
6.7	MALDI spectrum of evaporated ET	82
6.8	Suggested decomposition of ET	83
6.9	Optical micrograph of two ET samples	83
6.10	Optical micrograph of ET on glass	84
6.11	X-ray diffraction of ET on glass	85
6.12	X-ray diffraction of sample A5	87
6.13	X-ray diffraction of sample A6	88
6.14	Optical micrograph of Sample 1 of (ET)(TCNQ)	89
6.15	X-ray diffraction of Sample 1 of (ET)(TCNQ)	90
6.16	Optical micrograph of Sample 2 of (ET)(TCNQ)	91
6.17	X-ray diffraction of Sample 2	91
6.18	Optical micrographs of (ET)(TCNQ) samples	92
6.19	X-ray diffraction of sample 20.06.07	93
6.20	X-ray diffraction of Sample 18.10.07	94
6.21	Optical micrograph of the sample from 06.11.07	95
6.22	X-ray diffraction of sample 06.11.07	96
6.23	Optical micrograph of $ET_2Cu(NCS)_2$ - Sample 1	97
6.24	X-ray diffraction of $ET_2Cu(NCS)_2$ - Sample 1	98
7.1	Optical micrograph of (ET)(TCNQ) crystal	100
7.2	I(T) measurement	101
7.3	Optical micrographs of sample C1	102

7.4	Optical micrographs of sample C10	102
7.5	SIN junction in the semiconductor picture	103
7.6	Planar cross junction geometry	103
7.7	Modified junction geometry for semiconductors	104
7.8	Tunneling measurement setup	105
7.9	Optical micrograph of Al-AlO$_x$-CuPc-Au tunnel junction (T5)	107
7.10	Tunnel spectrum of an Al-AlOx-Nb contact	108
7.11	Tunnel spectrum of an Al-AlOx-CuPc-Nb contact	108
A.1	Sketch of the currently available stencil mask types	112
B.1	Wall coverage of an open cell after migration	114
B.2	Sketch of typical EBID geometry	115
B.3	Sketch of EBID geometry with beveled capillary	116
B.4	Thickness distribution for tilted substrates	117
B.5	Thickness distribution for tilted substrates	118
B.6	Thickness distribution for a beveled capillary	119

List of Tables

3.1	Table for the sample bake-out	33
5.1	Direct beam cut-off	67
5.2	Comparision of the different λ's	71
6.1	Co-evaporation of ET and TCNQ	86
7.1	Overview over the prepared tunnel contacts	106
A.1	List of all ports of the OMBD chamber	111
C.1	ET evaporation from graphite and quartz crucibles	121
C.2	Overview of the prepared samples of TCNQ	122
C.3	Evaporation of (ET)(TCNQ)	122
C.4	Co-evaporation of (ET)(TCNQ) and TCNQ	123

Introduction

Until the early 70's, it was believed that organic materials and polymers show interesting mechanical properties, such as softness and flexibility and offer new preparation methods such as printing and casting, but were generally regarded as insulating. Since the discovery of the first molecular metal TTF-TCNQ in 1973 [1] and of conductive polymers in 1977 [2] a new field of research has emerged.Suddenly, organic materials became interesting for electronic circuits creating the field of *molecular electronics*. With the availability of insulating, semiconducting and metallic organic layers the idea of all-organic circuits came up. The organics offer a unique combination of mechanical and electrical properties which power the idea to have cheap, easily processable and flexible electronics. This goal has partially been reached today with already available organic light emitting diodes (OLEDs)[1], organic field effect transistors (OFETs) [3] and organic solar cells[2]. All these devices have in common that they rely on the preparation of thin films of the used molecules.

A further advantage of organics is the large variety of substances as compared to inorganics and the possibility to tailor the molecules by means of organic synthesis. This feeds the dream to engineer materials to custom needs, once the the underlying principles of material properties are understood. Still, even though enormous efforts in research and development at universities and in industry have been performed, the fundamental properties of the materials are not fully understood. The precise nature of conduction in organic semiconductors and metals [1] or the origin of superconductivity [4] or other electronic phenomena is often unclear.

Charge transfer (CT) salts are a special class of compounds, constituting of a donor and an acceptor, between which a charge transfer occurs. For organic CT salts the partial ionic binding caused by the charge transfer is one reason for the interesting electronic properties of these compounds. As an example, more than 100 organic charge transfer superconductors have been discovered up to now [5], many of them based on the donor bisethylenedithiotetrathiafulvalene (BEDT-TTF), or short ET [6].

In this work charge transfer salts of ET are studied. Up to now semiconducting [7], metallic [8] and superconducting [9] CT salts of ET are known. The aim of this work was to prepare thin films of the materials with the ultimate goal to study their electronic properties with tunneling spectroscopy. Tunneling spectroscopy is a method originally invented for superconductors. There, the energy gap can directly be measured and information about the order parameter can be deduced. While today a large field of research deals with local tunneling

---
[1] http://www.samsungsdi.co.kr/contents/en/product/oled/oled.html
[2] http://www.konarka.com/

with an scanning tunneling microscope (STM), here the original method by Giaever [10, 11] was employed. For this method the availability of well-defined thin film layers is of particular importance and therefore, a necessity to search for optimized growth conditions. The first representative of interest was κ-ET$_2$Cu(NCS)$_2$, which is a superconductor below 10.4 K [9] and which has structural similarity to high temperature (High T_c) superconductors. Secondly, the class of (ET)(TCNQ) was studied.

Chapter 1 gives an introduction to charge transfer and charge transfer salts. In chapter 2 follows a description of the materials focused on in this thesis. Then construction and setup of a chamber for organic molecular beam deposition (OMBD) are described in chapter 3. The design of various effusion cells is presented in chapter 4 and their evaporation characteristics are compared to a numerical simulation in chapter 5. In chapter 6 the experiments on thin film preparation for the different materials are summarized. First measurements are presented in chapter 7, where the method of tunneling spectroscopy will be introduced shortly.

Chapter 1

Charge Transfer

1.1 Basic concepts

The definition of a charge transfer (CT) complex according to the international union of pure and applied chemistry is

> an electron-donor-electron-acceptor complex, characterized by electronic transition(s) to an excited state in which there is a partial transfer of electronic charge from the donor to the acceptor moiety [12].

One can express this as a reaction equation, with A denoting the acceptor and D the donor

$$D_m + A_n \rightarrow D_m^{+x} + A_n^{-x}, \qquad (1.1)$$

where m and n are integers, and x is the amount of charge transferred from donor to acceptor. The energy difference between the initial state and the final state can be written as

$$\Delta E_{CT} = I_0 - A_e - X \qquad (1.2)$$

where I_0 is the ionization energy of the donor, A_e the electron affinity of the acceptor and X summarizes the contributions of Coulomb energy, polarization energy and exchange energy [5]. A low ionization potential of the donor, together with a high electron affinity of the acceptor is favorable for the formation of a stable CT complex.

For organic molecules as donor and acceptor the energy levels can be calculated within the framework of a tight-binding model, where the molecular orbitals are derived from the original atomic orbitals of the constituents of the molecule. Here the highest occupied molecular orbital (HOMO) and the lowest unoccupied molecular orbital (LUMO) are of importance, and charge transfer occurs from the HOMO of the donor to the LUMO of the acceptor (cp. Fig. 1.1).

A charge transfer is a change in the electronic structure, which is accompanied by a change in the chemical structure. The CT complex usually has electronic properties that cannot be attributed to either of the reactants. First indications of this were that charge transfer complexes of two colorless reactants often display a strong color, exhibiting strong optical

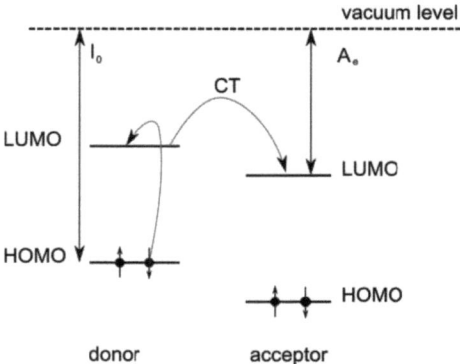

Figure 1.1: Energy diagram of charge transfer in organics
An electron on the donor is excited from HOMO to LUMO and can then be transferred to the lower lying LUMO of the acceptor. Indicated are also the ionization energy of the donor and the electron affinity of the acceptor. An energetically even more favorable situation exists, if the LUMO of the acceptor lies not only lower than the LUMO of the donor, but also lower than the HOMO, i.e. $A_e > I_0$.

absorption [13].

In this work, the focus lies on *intermolecular* charge transfer, where donor and acceptor are two different, separated molecules and the transfer therefore occurs *through space* in contrast to a *through bond* transfer. Since the transfer rate decreases exponentially with the distance between the reactants due to the overlap of electronic orbitals, the distance for a *through space* process is typically smaller than 20 Å [14].

With neutral donor and acceptor as the initial state and the charge transfer complex as the final state, one can describe the process in the framework of a double-well potential (Fig. 1.2), where the final state can either be reached via crossing the barrier, which costs an activation energy of E_A, or tunneling through the barrier. The coordinate q separating the two states, is the so-called reaction coordinate and is a measure for the change in configuration from reactants to product and may be e.g. bond length, bond angle, distance or nuclear arrangement. For complex molecules the reaction coordinate is usually a *collective coordinate* accounting for the nuclear rearrangement during the process and all relevant degrees of freedom. Apparently, the difficulty of defining the reaction coordinate increases with the size of the involved molecules. In the general case the simple energy curve in Fig. 1.2 has to be substituted by a complex potential energy surface.

According to Arrhenius' law one expects the rate constant for an electron transfer from the left minimum in Fig. 1.2 to the right minimum to be

$$k_{CT} = \nu \cdot e^{\frac{-E_A}{k_B T}}, \tag{1.3}$$

where E_A is the activation energy, T the temperature, k_B Boltzmann's constant and ν is a frequency factor. Marcus has indeed derived this result for a simplified model of electron

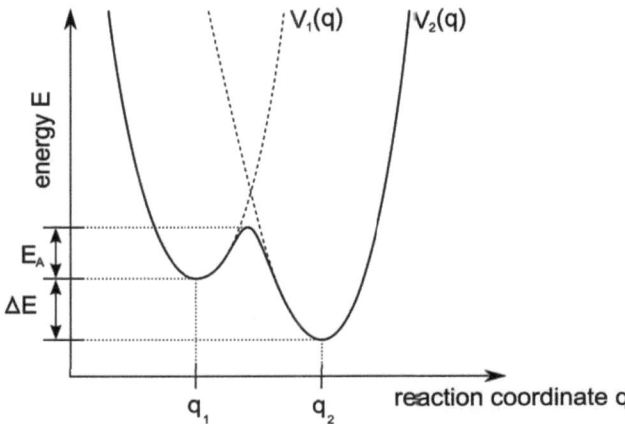

Figure 1.2: Schematic potential energy curve for a CT reaction
The left minimum describes the initial state before the transfer occurs, the right minimum describes the final state after the charge transfer. The height of the potential well is the activation energy necessary for the reaction and the coordinate q is the reaction coordinate.

transfer in solution [15]. The frequency term strongly depends on the precise nature of the examined problem and also the activation energy is not easily accessible in practice. In general, charge transfer remains a quantum mechanical many body problem, where most solutions up to now are based on calculations for simplified model systems.

1.2 CT in the gas phase/on a substrate

A major area of research concerns charge transfer in solution, where polarization of the solvent molecules plays an important role, as well as diffusion of the constituents in the medium. A second area is internal charge transfer, where the charge transfer is often assisted by molecular bridges between a donor and acceptor part of a macromolecular entity.

In contrast, the interest here lies in charge transfer in the gas phase or on a substrate (S). The employed preparation method is *co-evaporation*. This means the two molecules D and A are evaporated separately under ultra-high vacuum conditions with the respective molecular beams overlapping at the substrate.

One crucial question is, whether the charge transfer in this experimental situation predominantly occurs in the gas phase or on the substrate. As an estimate the time for an encounter of two molecules in the gas phase can be determined. The mean thermal velocity of the evaporated molecules is

$$\bar{v} = \sqrt{\frac{8R \cdot T}{\pi M}}, \qquad (1.4)$$

where R is the universal gas constant, M is the molecular weight and T is the temperature. If one assumes an evaporation temperature of about 135 °C, a molecular weight of 200 − 400 g/mol (TCNQ = 204.2 g/mol, ET 384.7 g/mol) and a sample to source distance of 250 mm, which reflects the experimental circumstances, the time of flight is in the range of a few milliseconds. The effective time is even shorter if only the region of the overlap of the two molecular beams is considered. The time for a possible encounter in the gas phase has to be compared to typical preparation times of several hours for which the molecules diffuse on the substrate. Furthermore, as mentioned earlier, charge transfer typically occurs at distances of about 20 Å or closer. Considering the short time available and the close distance necessary, the probability of a charge transfer in the gas phase is small. Therefore the suggestion is, that CT occurs predominantly on the substrate.

As a consequence one has to consider e.g. the following parameters influencing the process: substrate temperature, substrate-molecule interaction for both donor and acceptor, mobility of the molecules on the surface and polarizability of the substrate. Some of these parameters are similar to considerations for CT in solution, with the difference that the substrate is geometrically anisotropic, being a mostly rigid flat surface, compared to a rather isotropic, surrounding solvent.

The charge transfer is governed by the complex donor-acceptor interaction (D/A), but here additionally substrate-molecule interactions (S/A and S/D) have to be taken into account. The interactions D/A, S/A and S/D cannot be treated independently, since each interaction changes the electronic structure of the involved partners, which in turn is the basis for the calculation of the interactions.

The molecule-substrate interaction can be summarized under the term *interface* interactions. Complementary to the considerations of charge transfer some relevant interactions concerning the interface are

- **Coulomb interaction**, most prominent for materials with ionic character and high formal charges.

- **Elastic interactions**, caused by lattice mismatch between the molecular crystal and the substrate. They may lead to defects or dislocations and influence the growth mode of the thin film.

- **Electron transfer across the interface**, depends on the electronegativity of the atoms at the interface. It occurs preferentially at metal/non-metal interfaces, where the transfer may lead to interface charges or gap states in the non-metal.

- **Pauli repulsion**, important for an interface of two closed-shell materials.

- **Image charge interactions**, at the interface of a metal with an ionic solid [16].

One expects a different behavior for metallic, semiconducting and insulating substrates, since the relevant interactions and their respective strength vary strongly. A general problem is to identify and calculate the relevant interactions. For the calculation of a given interface a detailed knowledge of the structural and electronic properties is necessary. This information is usually not readily available.

For organic materials in particular the high structural variability adds complexity to the problem and due to the many possible molecular orientations it is nearly impossible to predict a favorable orientation of a molecule on the substrate.

1.3 Physical Properties

For this work transport properties such as metallic conduction and superconductivity are of interest, which have both been observed in charge transfer salts [1, 9]. To obtain a conducting compound, free charge carriers are necessary and for the formation of a conduction band the carriers have to delocalize throughout the crystal. In organics this is usually realized by π-*stacking*: The molecules arrange themselves with an overlap of the π-orbitals of adjacent molecules. The extent of overlap determines the band width and can e.g. be influenced by applying pressure. Among the organic materials charge transfer salts exhibit the largest band width of $0.2 - 0.5$ eV compared to a typical value of 0.1 eV for organics. The reason for this is the contribution of ionic binding due to the CT.

Moreover, the CT is often the source of the free carriers. This means that the number of carriers is usually fixed according to the amount of charge transferred and therefore independent of temperature. A typical charge transfer in organics is in the range of 0.5 to 1 electrons per unit cell. The charge carrier concentration is accordingly the amount of charge transferred per molecular entity divided by the size of the unit cell. Typically, charge carrier densities in the order of 10^{21} cm^{-3} are expected. In comparison, copper has a charge carrier density of $8.5 \cdot 10^{22}$ cm^{-3} [5].

A common feature of organic charge transfer salts is the anisotropy of properties, which is caused by the anisotropic building blocks of the crystal. In particular, systems of reduced dimensionality are found, where a 2D or 1D character is caused by a sheet-like [9] or chain-like stacking [1] of the molecules in the crystal. Furthermore organic crystals are usually soft lattices due to the weak Van der Waals interaction which makes them especially receptive for the application of pressure [17].

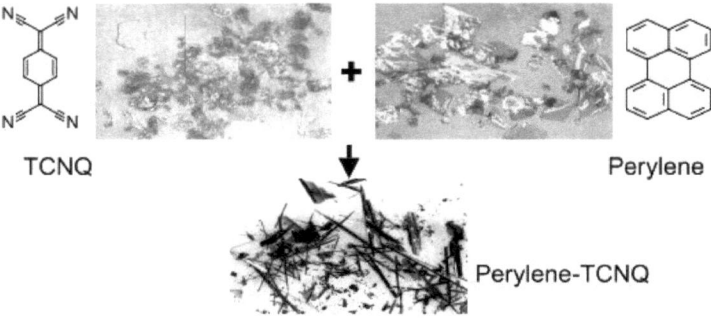

Figure 1.3: CT salt formation of Perylene-TCNQ
The source materials Perylene and TCNQ with the respective chemical structure are shown in the upper photographs. The lower photograph shows the prepared charge transfer salt Perylene-TCNQ. The pictures are taken from [18].

1.4 Examples

In the following selected examples of charge transfer in organics are presented regarding the properties and preparation methods of interest for this work.

1.4.1 CT in the gas phase

The formation of the charge transfer salt Perylene-TCNQ was observed by Zeis [18] after the evaporation of the two constituents. The charge transfer crystals were prepared with physical vapor transport (PVT) in a flow of inert gas. Photographs of the source materials Perylene and TCNQ are shown in the upper pictures of Fig. 1.3, and the resulting charge transfer salt in the lower picture. As a clear indication for the CT a drastic color change from originally yellow to dark green accompanies the charge transfer process. Three growth zones could be identified during the experiment separating Perylene, TCNQ and the CT salt. Due to the occurrence of three distinct zones the assumption is that the CT process happened in the gas phase. This does not contradict the statement that under UHV conditions a CT on the substrate is more likely. In PVT a transport gas is present facilitating collisions of the evaporated molecules. This is a very different experimental situation, but it proves that a CT in the gas phase is in principle possible.

1.4.2 Molecular metals

A prominent example of an organic charge transfer salt is (TTF)(TCNQ), which was synthesized in 1973 by Ferraris et al. [1]. It was the first so-called molecular metal with a maximum conductivity of $1.47 \cdot 10^4$ $\Omega^{-1}\text{cm}^{-1}$ at 66 K. It consists of separated TTF and TCNQ stacks, forming one-dimensional chains, with preferred conduction along the chains. It has been exten-

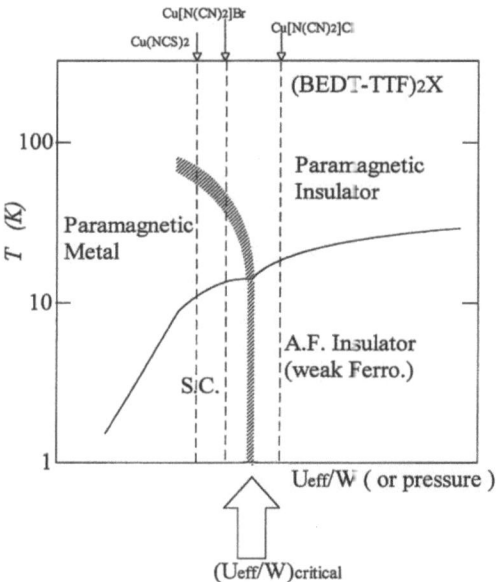

Figure 1.4: Generic phase diagram of κ-(ET)X_2
Kanoda proposed this generic phase diagram for the κ-(ET)X_2 salts, based on NMR, magnetic susceptibility and specific heat measurements [20]. With increasing pressure the orbital overlap is increased and accordingly the band width W, whereas U_{eff}, the on-site Coulomb interaction is mainly unaffected. The shaded region in the phase diagram marks the position of a Mott transition, yielding anomalous behavior in the transition region. A.F. denotes the antiferromagnetic phase and S.C. the superconducting phase.

sively studied as a prototype of an one-dimensional metal. The amount of charge transferred per TTF molecule has been determined by x-ray photoemission to 0.59 electrons [19].

This example shows also a general rule regarding the stacking of the molecules. A donor acceptor arrangement in segregated stacks results in a semiconducting or metallic compound, whereas a mixed-stacked arrangement results in a semiconducting or insulating compound.

1.4.3 Organic superconductors

The κ-ET-salts show great similarity to the high temperature (high-T_c) superconductors, discovered by Bednorz and Müller [21]. This similarity manifests in a structural similarity and in an analogy of electronic properties. From a structural point of view, the conducting ET-layers resemble the sheet-like conducting copper-oxide planes of the high-T_c superconductors. It has been shown that the superconducting transition in the high-T_c superconductors can be shifted or suppressed by means of doping. For the organics a similar behavior can be observed by means of, e.g., pressure.

A generic phase diagram for the ET-salts is shown in Fig. 1.4, which also indicates the effect of pressure [20]. With increasing pressure the orbital overlap is increased and accordingly the band width W, whereas U_eff, the on-site Coulomb interaction is mainly unaffected. Therefore a system can be tuned along the x-axis of the phase diagram, by applying pressure. There are two ways, how this can be realized: as external pressure, or as internal (chemical) pressure, due to the exchange of atoms. An example for this, depicted in the phase diagram, is the substitution of Cl with Br, for the compound Cu[N(CN)$_2$]Cl, where Br has a larger radius.

If the phase diagrams for high-Tc superconductors and ET-salts is compared (cp. Fig. 1.5), several of the similarities become obvious. Both show a antiferromagnetic area, close to a superconducting dome. Above the superconducting area a metallic state is realized, which undergoes a transition to a strange metal (high-T_c) or insulator (ET X_2). This is indicated by the dotted line within the phase diagram. The high-T_c phase diagram is strongly simplified and shows in fact more features depending on the specific compound.

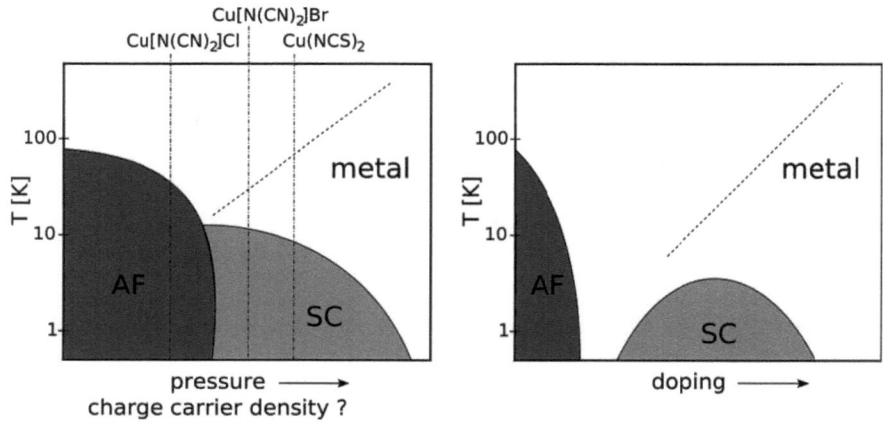

Figure 1.5: Generic phase diagram of (ET)X_2 compared to high-T$_c$s
Shown are two very simplified phase diagrams for (ET)X_2 compared to high-T$_c$ superconductors. These serve to strengthen the similarity of the two systems in terms of electronic properties. Clearly, the superconducting region in vicinity of an antiferromagnetic region is a first feature. Secondly, there is a transition from a normal metal state to a strange metal, metallic or insulating state, depending on the specific compound. This transition is symbolized by the dotted line. In this region anomalous behavior is observed.

Apart from tuning the superconducting transition through doping as for the high-T$_c$ superconductors or through the application of pressure for the organics, one can also think about a direct variation of the charge carrier density, e.g. using the electrostatic field effect.

For both types of superconductors the occurrence of superconductivity is not fully understood up to now. Both are considered "unconventional" superconductors, since the order parameter shows d-wave symmetry.

Summing up, CT salts are a promising class of organic compounds, which show interesting physical behavior and can serve as model systems for low-dimensional materials and high-T_c superconductors.

Chapter 2

Materials and Methods

2.1 Materials

The organic materials used for the preparation are shortly presented. The materials were chosen according to two main aspects, namely

- electronic properties and
- experimental handling.

Regarding the first aspect, the group of ET salts offers a variety of interesting compounds, including the subgroup of superconducting $\kappa - (BEDT - TTF)_2X$ salts. The second aspect poses more severe restrictions on the choice of materials. The preparation method for the CT thin films employed in this work is the co-evaporation. The prerequisite for this method is that donor and acceptor are separately available as stable compounds and that they are sublimable. Unfortunately information on sublimability of the molecules is often either not available or inconclusive.

2.1.1 ET

The electron donor ET, $C_{10}H_8S_8$, molecular mass $M_{et} = 384.68$ g/mol, is a derivative of the organic donor tetrathiafulvalene (TTF). The crystals display a bright orange color. ET is a versatile building block for charge transfer complexes, because of various properties:

- It has (in CT complexes) a planar shape, which is favorable for the π-stacking of the molecule. The molecular structure is shown in Fig. 2.1.

- The additional sulfur promotes a two dimensional cross-linking and increases dimensionality as compared to TTF, which preferably forms one dimensional chains as in (TTF)(TCNQ).

For this work it is of importance that ET is sublimable at temperatures from 80 °C [22] to 200 °C [23], in contrast to TTF, which has a significant vapor pressure already at room temperature. ET has already been used in OLEDs for n-type doping [24] and has been shown to form microcrystalline thin films on KCl [23].

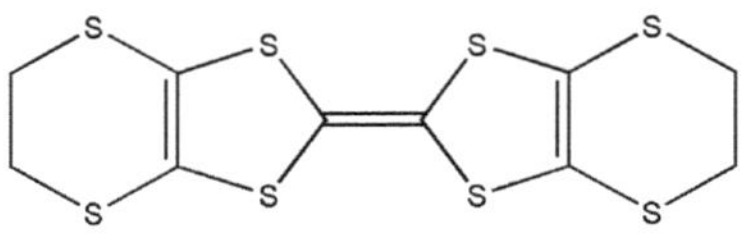

Figure 2.1: Molecular structure of BEDT-TTF

Structure ET crystallizes in the monoclinic space group $P2_1/c$ [25]. The molecules form pairs in the crystal. The neutral ET is non-planar due to the intermolecular contacts formed by the sulfur atoms. Structural analysis of CT salts of ET have however shown, that the charged ET in the CT crystal has a planar shape [26].

2.1.2 Cu(NCS)$_2$

Cu(II)-thiocyanate is a brown solid. The molecular mass of the compound is $M_{\text{cuncs}} = 179.71$ g/mol. It is needed as a constituent of ET$_2$Cu(NCS)$_2$, described in the next section.

The substance is highly hygroscopic and decomposes to Cu(I)-thiocyanate in the presence of water. It also thermally decomposes to Cu(I)-thiocyanate upon heating to temperatures of 170 °C- 180 °C in air, where the decomposition reaction again involves water [27, 28]. A reference for vacuum sublimation of the compound was not found.

2.1.3 ET$_2$Cu(NCS)$_2$

ET$_2$Cu(NCS)$_2$ was first synthesized in 1988 by Uruyama et al. [9] and is an ambient pressure organic superconductor with a T_c of 10.4 K.

Structure ET$_2$Cu(NCS)$_2$ crystallizes in the κ-phase, in the space group $P2/1$. Two ET-molecules form a face-to-face dimer and the dimers form a conductive ET-layer. The ET-planes are separated by planes of Cu(NCS)$_2$ forming an insulating polymer layer [29, 30]. The ethylene end groups of the ET-molecules show conformational disorder at room temperature and freeze out below 100 K [29, 31].

Electronic properties The copper has been shown to be Cu(I), which corresponds to a charge transfer of one electron per ET dimer, resulting in a half-filled band [32].

The nature of superconductivity of is still unclear. Evidence for d-wave pairing symmetry has been observed in scanning tunneling spectroscopy measurements [33].

2.1.4 TCNQ

The electron acceptor tetracyanoquinodimethane (TCNQ), $C_{12}H_4N_4$, molecular mass $M_{\text{tcnq}} = 204.2$ g/mol, is a yellow solid [34]. The molecule is planar. The molecular structure is shown in Fig. 2.2. An interesting property of TCNQ is that the acceptor strength can be modified by

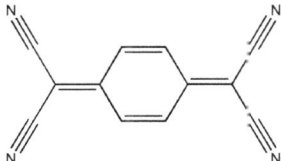

Figure 2.2: Molecular structure of TCNQ

substitution. Fluorinated derivatives F_nTCNQ with $n = 1, 2, 4$ are available. In the case of F_4-TCNQ the four hydrogen atoms at the carbon ring are substituted by fluor atoms. This allows in principle for the preparation of isostructural compounds with gradually varying electronic properties. Experiments with ET and F_n-TCNQ, with $n = 1, 2$ have however resulted in a different crystal structure compared to (ET)(TCNQ) [35]. Also, TCNQ substitued with chlorine and methyle groups has already been studied in the context of CT salts [36].

2.1.5 (ET)(TCNQ)

The charge transfer salt (ET)(TCNQ) is known to form three structural variants: the monoclinic phase, the β'-phase and the β''-phase.

Monoclinic phase

The monoclinic phase consists of a mixed-stacked configuration of ET and TCNQ. The compound is insulating, with a resistivity of about $\rho = 10^6 \, \Omega \cdot \text{cm}$ [26].

β'-phase

The β'-phase has a triclinic crystal structure with ET planes separated by TCNQ columns. The TCNQ columns are stacked face-to-face along the c-axis, the ET molecules are arranged side-by-side. The compound is semiconducting and a metal-insulator transition at 330 K at ambient pressure has been observed [7]. Iwasa et al. have studied the transition depending on applied pressure [37] and found that the transition temperature shifts with pressure as shown in Fig. 2.3. The corresponding phase diagram is shown in Fig. 2.4. More recent experiments of Miyashita et al. result in a different phase diagram [38]. The phase diagram proposed by Miyashita is also shown in Fig. 2.4 for comparison. The difference in both models is the slope of the phase boundary above 0.5 GPa. Miyashita claims, in contrast to Iwasa, that the metal insulator transition can be fully suppressed by the application of pressure, resulting in a metallic state down to 0 K.

β''-phase

The β''-phase of (ET)(TCNQ) was first found in 2003 by Yamamoto et al. [8]. The crystal structure is triclinic with ET layers separated by one dimensional TCNQ stacks. The structure

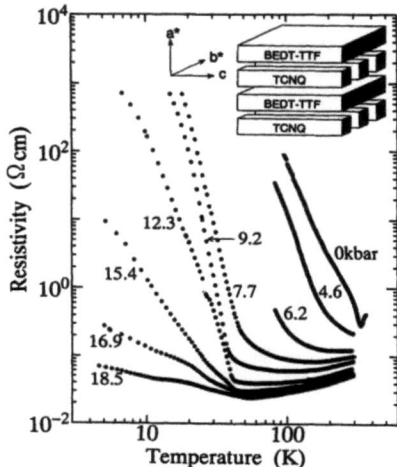

Figure 2.3: Metal-insulator transition of (ET)(TCNQ), β'-phase
The graph from Iwasa et al. [37] shows the temperature-dependence of the resistivity of (ET)(TCNQ), β'-phase under pressure. Several measurements between 0 kbar and 18.5 kbar have been performed. In the curve at ambient pressure a minimum in the resistivity at around 300 K marks the point of the metal-insulator transition. This transition is suppressed under application of pressure.

is shown in Fig. 2.5. The charge transfer is estimated to $0.74e$ per ET molecule. The compound shows metallic behavior with several anomalies at 20 K, 80 K and 170 K. This structure variant has solely been obtained using a special catalyst, which is assumed to act as a template for crystal growth. The electronic properties were studied with respect to resistivity, magnetic susceptibility, electron spin resonance (ESR) [39] and magnetooptical properties [40], but the results are not conclusive up to now.

First field-effect transistors of (ET)(TCNQ) have been prepared by drop-casting a solution of ET and TCNQ and chloroform on a n-type doped silicon substrate; bipolar behavior was observed [41, 42].

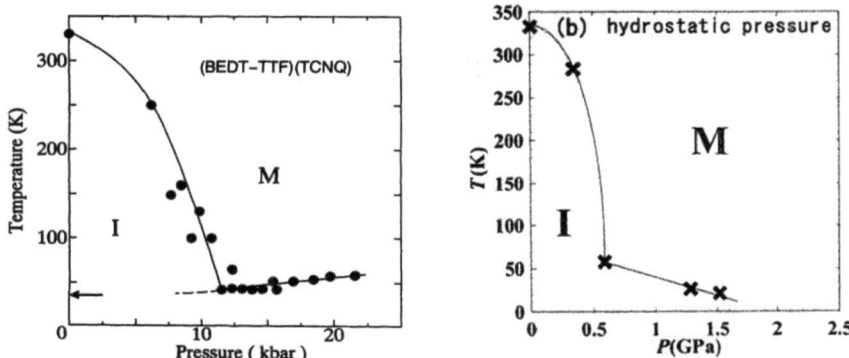

Figure 2.4: Proposed phase diagrams for (ET)(TCNQ), β'-phase
On the left hand side the phase diagram as proposed by Iwasa et al. is shown [37]. On the right hand side the proposal of Miyashita et al. [38] is shown in comparison. The difference between the two is the slope of the phase boundary with increasing pressure. Above 0.5 GPa Miyashita finds a negative slope, allowing to fully suppress the metal-insulator transistion.

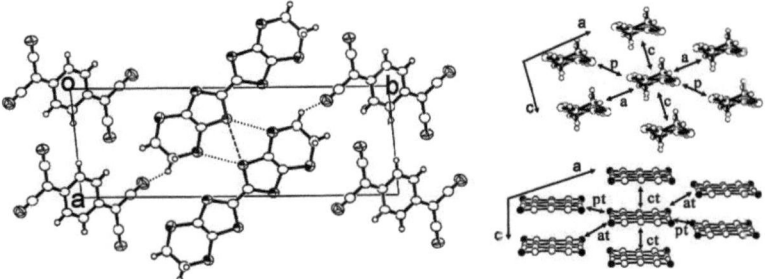

Figure 2.5: Crystal structure of (ET)(TCNQ), β''-phase
On the left: view along the c-axis, upper right: end-on projection of the donor layer and lower right: projection of the acceptor layer [8].

2.2 Characterization methods

Three characterization methods are shortly introduced, regarding structure and composition of thin films and single crystals. Among them, X-ray diffraction was the most frequently applied method.

2.2.1 X-ray diffraction

The structural analysis of all thin film and powder samples was performed with X-ray diffraction on a Bruker D8 diffractometer. A photograph of the setup is shown in Fig. 2.6. The diffractometer has a parallel beam optic and is especially suited for the measurement of thin films, including the possibility for X-ray reflectometry for thickness measurements and a grazing incidence setup for in-plane measurements.

For the organic thin film samples only normal diffraction was employed due to the fact, that the samples were usually microcrystalline and not forming a closed layer. The measurements were typically running for several hours, summing up repeated scans of short duration. Only then sufficient statistics could be obtained.

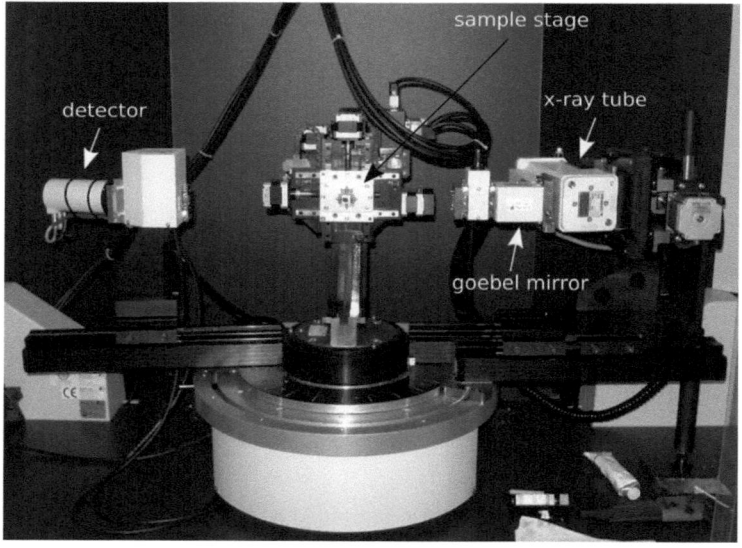

Figure 2.6: Bruker D8 diffractometer

On the right hand side the X-ray tube is mounted. In front of the tube the goebel mirror is attached and before and after the mirror slits can be inserted into the beam. In the middle, the sample is fixed on an x-, y-, z- stage. The whole sample stage is mounted on three circles, ϕ, χ and ω. The detector is mounted on the 2θ-circle. A variable slit, or a germanium analyzer crystal can be used on the detector side. In this thesis mainly $\omega/2\theta$-scans are performed, with all other parameters fixed.

Powder samples were also studied, although the diffractometer setup is not typical for powder measurements, due to the parallel beam optics. To verify the results, comparative measurements were performed on a Siemens D 500 with Bragg-Brentano geometry and showed excellent agreement. The crystallites were powdered with a mortar and spread onto a microscope slide prepared with a thin layer of vacuum grease. The typical measurement settings on the D8 were large slits on the X-ray source side (slits 1 mm and a square slit 5×5 mm^2) and a variable slit on the detector side (setting V4).

2.2.2 MALDI

MALDI is matrix assisted laser desorption ionization, a type of mass spectrometry first described in [43, 44]. It is especially suited for the study of large organic or bio-molecules. The material to be studied is mixed with a solvent and a matrix and pipetted onto a sample holder. After the solvent has evaporated the holder is inserted into a mass spectrometer, where the matrix is sublimed with a laser pulse, shot onto the prepared area. The material is thereby ionized and detected with the spectrometer, together with fragments of the matrix. The role of the matrix is to protect the material from the direct laser beam and to facilitate evaporation and ionization.

For the study of thin film samples, the deposited material was scratch from the substrate and mixed with solvent and matrix. This provided only small amounts of material allowing only for the preparation of one or two MALDI samples per thin film sample. Still, the amount was so little, that it was difficult to detect the source material with the spectrometer.

The spectrometer used for the experiments is a FISONS Instruments VG TofSpec and the measurements were carried out in the chemistry department.

2.2.3 EDX

Energy dispersive X-ray spectroscopy (EDX) is a method for element analysis. The EDX setup is most commonly included in a scanning electron microscope (SEM), which is therefore equipped with an X-ray detector. Sample atoms are excited by the bombardment with electrons from the electron beam. Upon de-excitation X-ray photons with characteristic energies according to the electronic transition are emitted. These are detected and provide an energy spectrum characteristic for the respective element. The analysis of the spectra provides quantitative information on composition for bulk material. For the investigation of thin films special corrections have to be taken into account.

The energy of the X-ray photons scales with the atomic mass, *light* elements are therefore difficult to detect. A special situation is given for carbon. Carbon is constantly deposited onto the sample from the residual gas by cracking of carbon-containing molecules through the electron beam. Accordingly, carbon is especially unsuited for EDX detection. Hence, organic molecules, build mainly from carbon and other light elements are not very suitable for element analysis with EDX.

For the organic molecules in this thesis there is a particular situation, namely, that the molecules each have an exclusive *marker* element. For TCNQ this marker is nitrogen, for ET it is sulfur. Both elements could be resolved with the existing setup. By only concentrating on the sulfur and nitrogen content in the samples, the existence and ratio of [ET] and [TCNQ] could be determined for various samples. EDX measurements were performed by Christina Grimm on a FEI Nova NanoLab 600 scanning electron microscope equipped with an EDAX Genesis 2000 system.

Chapter 3

Molecular beam epitaxy system

Several different methods are known for the preparation of organic thin films, e.g. chemical vapor deposition, spin coating, ink-jet printing and molecular beam epitaxy (MBE). These techniques have individual advantages and disadvantages and the decision for any of them depends strongly on the requirements one imposes on the preparation process.

Molecular beam epitaxy is the thermal evaporation of source material onto a single crystalline substrate. It has the advantages that

- it is a clean process, because it is carried out under ultra high vacuum (UHV) conditions, corresponding to a pressure below 10^{-7} mbar,

- film thickness, growth rate and deposition temperature can be controlled accurately and

- it is a solvent-free method and can also be used for non-soluble molecules, e.g. pigments.

The main disadvantage is the possible decomposition of organic molecules during evaporation.

As an important part of this work an organic molecular beam deposition (OMBD) chamber was designed, set up and connected to the existing metal-MBE (MMBE) chamber. An overview of the five-chamber system is shown in Fig. 3.1. In the following functionality and special features of the system are presented with focus on the OMBD chamber.

3.1 MMBE-chamber and sample transfer

MMBE chamber and sample transfer system were purchased from BesTec GmbH, Berlin. The sample transfer is accomplished with sample carriers onto which substrates with a diameter of up to 2 inches can be fixed.

The two-piece transfer chamber provides in total four ports, where chambers can be connected to the system. A track traverses the transfer chamber. A trolley with a revolving cable moves on the track driven by a rotary feedthrough and distributes the sample carriers.

The MMBE chamber, delivered with the transfer system, is equipped with three standard effusion cells (two high temperature effusion cells HTC-40-10-SH by CreaTec and one single

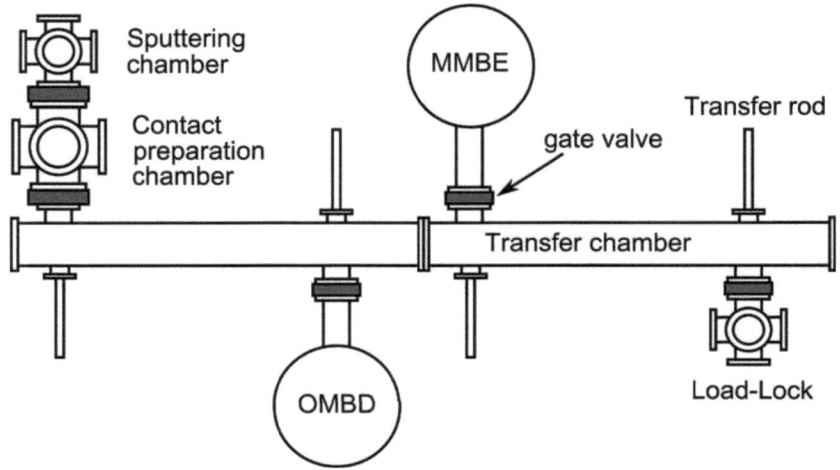

Figure 3.1: Sketch of the five-chamber MBE system
The system consists of a two-piece transfer chamber with a total of four ports for individual chambers with associated transfer rods. The ports are occupied by a load-lock, the MMBE chamber, the OMBD chamber and a two-chamber system of contact preparation chamber and sputtering chamber. All chambers can be operated independently due to individual pumping systems and gate valves in between the chambers.

filament effusion cell SFC-40-10-SH) providing Indium, Cobalt and Cerium as well as one electron beam evaporator together with a quartz microbalance (QMB) currently used for Niobium. The sample manipulator is capable of temperatures of up to 1100 °C and has an azimuthal sample rotation. Structural characterization tools available in this chamber are low energy electron diffraction (LEED) and reflection high energy electron diffraction (RHEED). A mass spectrometer is included to study the residual gas composition.

3.2 Load-Lock

In the load-lock three sample stages are mounted to a rotary feedthrough. The load-lock is equipped with one 250 W tungsten filament to bake the samples consecutively up to 400 °C before transfer into the main chamber system. The rotary feedthrough has been motorized and can be operated automatically to switch the sample carriers during bake-out. The bake-out temperature has been calibrated with a type-K thermocouple attached to a sample carrier. The results are shown in Table 3.1.

P [W]	T [°C]
80	190
100	225
120	260
170	350
200	395
210	405

Table 3.1: Table for the sample bake-out

The tungsten filament for sample bake-out was calibrated using a type-K thermocouple mounted on a sample carrier instead of the substrate. The bake-out temperature should be chosen according to the substrate material used.

3.3 OMBD-chamber

The chamber was manufactured by VTS-CreaTec GmbH according to our specifications. It consists of a cylindrical body of height 625.5 mm and diameter 400 mm with a total of 30 flanges including the CF 400 top flange. An overview over the existing flanges and their designated function is given in Appendix A.1. Position and tilt angles of the individual flanges were assigned according to the intended use. A 3D POV-Ray[1] model was drafted to simulate the field of vision for the view ports and perform collision checks for the internal fittings, e.g. effusion cells, sample manipulator and LEED optics. Two side views of the 3D model are shown in Fig. 3.2.

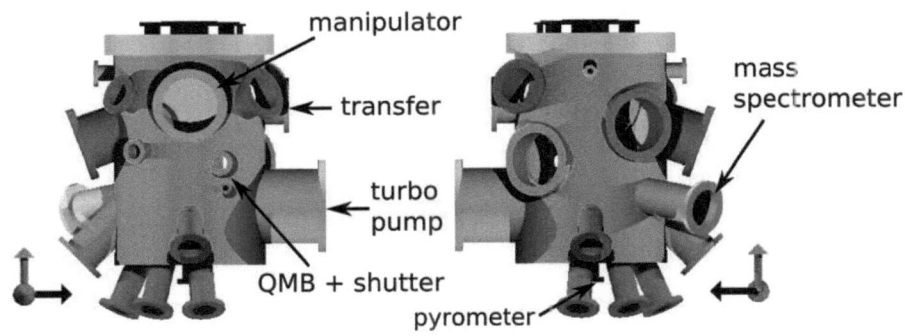

Figure 3.2: 3D model of the main chamber
Ports are assigned for: 6 effusion cells, view ports, LEED, cylindrical mirror analyzer, pressure gauges, RHEED electron gun/screen

3.3.1 Sample manipulator and cooling mechanism

The sample manipulator had to be adapted to the existing transfer mechanism and sample carriers from the BesTec system. Furthermore, the requirements for an organic deposition chamber included temperatures of up to at least 200 °C, cooling with liquid nitrogen and an azimuthal rotation to enable LEED and RHEED.

---
[1]Persistance of Vision Raytracer, http://www.povray.org/

The most effective heating method in this temperature range is thermal conduction. The heater is therefore a massive copper block with an integrated tungsten resistive heater. During sample transfer the heater is pressed to the back of the sample carrier with metal springs. For good thermal connection new sample carriers were made from copper. For temperature measurement a type-K thermocouple is clamped to the heater. The measurement takes place 1 mm away from the contact area of manipulator and carrier. The heating has been tested to about 400 °C.

The cooling assembly is mounted on a separate flange and can be attached to and detached from the manipulator. It consists of a massive copper block with a milled-into spiral structure. The block is encased in a stainless steel container with inflow and outflow for liquid nitrogen. With this design the passage for the liquid nitrogen in the container is as long as possible so that an efficient cooling takes place. Cooling and manipulator both have a copper bar that can be brought into contact to connect the cooling. The cooling container is therefore mounted on a linear travel. It should be noted that the linear travel has to be adjusted during cooling, because the pipes shorten due to thermal expansion and the cooling may therefore detach itself from the manipulator. The lowest temperature achieved was -57 °C after about 110 min of cooling with all heat sources (effusion cells, pressure gauges) turned off. Liquid nitrogen was constantly filled into a funnel connected to the inlet. The outlet was pumped with a rotary vane pump. The temperature was measured with the manipulator thermocouple. With simultaneous evaporation at 136 °C in a distance of about 15 mm for 120 min only -14.7 °C were reached.

The azimuthal rotation is realized with a AML vacuum motor. The motion is translated with a pair of bevel gears. The rotation is not continuous, but $\pm 180°$ from the transfer position due to the electrical connections of the manipulator. A mechanical fixture prevents undesired rotation during sample transfer.

3.3.2 Characterization tools

Apart from pressure gauges, further tools for in-situ characterization are provided. A quadrupol mass spectrometer QMA 120 with QME 112 and control unit QMG 112 from Balzers for the study of the residual gas composition is attached. It is not possible to monitor decomposition of molecules or charge transfer processes in the gas phase, since only atomic masses up to 200 u can be detected. Furthermore, there is a VG Scientific rear-view LEED optics RVL 640 with 747A power supply. For rate monitoring a QMB with a separate shutter is available, which is positioned in such a way that it is visible from each effusion cell port.

3.4 Sputtering & contact preparation chamber

Since only one free port of the transfer tube is remaining, sputtering chamber and contact preparation chamber are build back-to-back, with the contact preparation chamber situated closer to the transfer tube. This implies that the sample manipulator of the contact preparation chamber has to be retractable to allow access to the sputtering chamber. This is realized by a

Figure 3.3: Sample carrier with mounted stencil mask
The mask is fixated with two metal springs on the carrier. The mask holds four small templates of the desired structure. Accordingly, always four devices are prepared on a 10×10 mm^2 substrate.

linear travel of 25 cm length together with a straight couple that hosts the manipulator in the retracted position. Both chambers can be operated separately due to independent pumping systems and a gate valve between them.

3.4.1 Contact preparation chamber

The contact preparation chamber hosts a stencil mask system for device preparation and a wobble stick for the manipulation of the masks. Device preparation is carried out by consecutive evaporation steps employing different stencil masks to define the respective sample structures. The masks can therefore be mounted onto the sample carrier and transported with the sample into any of the other chambers. A sample carrier with mounted mask is shown in Fig. 3.3. This system of portable masks was favored over fixed masks in the individual chambers for its flexibility. Masks can easily be exchanged, modified or repaired after transport to the load-lock. A total of seven masks can be stored in the contact preparation chamber in a retainer. A list of the existing mask types is given in Appendix A.2. Two ports for effusion cells equipped with cooling shrouds are provided, one of which is equipped with a single filament cell for gold. The evaporation of gold is automated and is most conveniently operated over night. For details on time/thickness, refer to the thesis of Florian Roth [45]. Optimum position of the sample manipulator with respect to the gold effusion cell is 300° on the integrated scale.

The sample manipulator consists of a simple sample holder without cooling, heating or temperature control, since the gold is only used as contact material. The manipulator shaft is reinforced by a triangular arrangement of springs, to stabilize it over the total length of about 50 cm. Due to the limited space no sample shutter is available.

3.4.2 Sputtering chamber

The sputtering chamber is based on a CF-100 six-way cross. It is equipped with two sputter cathodes, one home-build cathode [46] and one *stilleto series magnetron sputtering source* from AJA International. The home-build cathode is used for aluminum sputtering, the other cathode for sputtering of $SrTiO_3$. Both cathodes are operated with a Hüttinger PFG 600 RF generator.

Additionally, a plasma oxidation stage was build to prepare an oxide layer on sputtered aluminum films for tunneling measurements (Fig. 3.4). An aluminum cylinder of 10 mm diameter is connected to a DC voltage source. An outer half-cylinder is added to compose an equipotential surface with the sample. Applying a voltage of about 500 V at an oxygen pressure of 0.4 mbar a plasma discharge can be ignited.

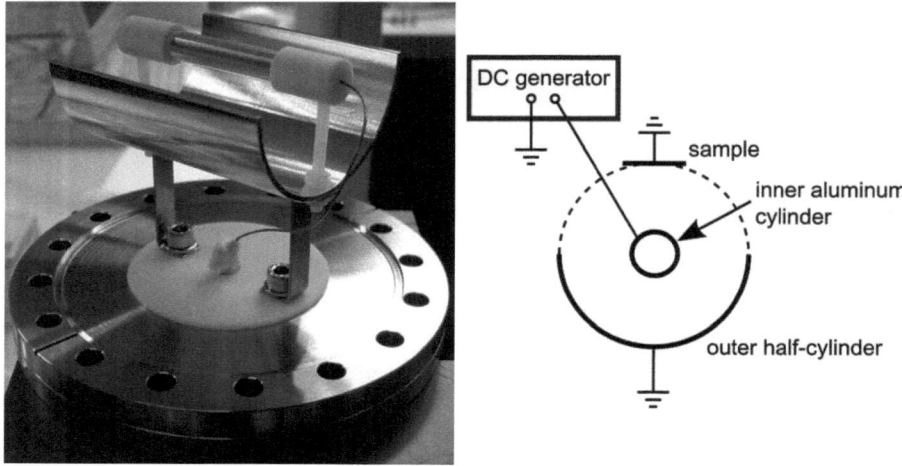

Figure 3.4: Plasma stage - photograph and sketch
The central aluminum cylinder is connected to an electrical feedthrough and isolated from the flange. The outer half-cylinder and the sample are electrically grounded through the chamber body and approximate a cylindrical equipotential surface. Applying a voltage of about 500 V at an oxygen pressure of 0.4 mbar a plasma discharge can be ignited.

The sputtering sample manipulator is capable of temperatures of up to 900 °C and is therefore water-cooled. To enable operation in oxygen atmosphere all hot metal parts of the manipulator consist of Inconel[2]. The resistive heating wire is Kanthal[3], which develops a protective aluminum oxide surface layer during the first operation in oxygen atmosphere. Details on the manipulator can be found in [45].

---
[2]nickel-based alloy, registered trademark of Special Metals Corporation
[3]iron/chromium alloy, http://www.kanthal.com/

The chamber has a gas-inlet system with two ports, equipped with oxygen and argon. The gas inlet is a two step design with gas flow controllers from Alicat Scientific regulating the flow from the gas bottles into a small volume outside the main vacuum chamber. This intermediate volume is connected to the main chamber via an UHV needle valve. This design has several advantages: The UHV valves allow for a perfect isolation of gas-inlet and chamber, but do not have reproducible opening states. The flowmeters allow for a reproducible gas flow with active regulation and allow for defined gas mixtures, but can not handle the total pressure difference, from 1.5 bar at the inlet to 10^{-4} mbar at the outlet. With this double design all advantages can be combined, eliminating the individual shortcomings.

3.5 Small MBE chamber

The first thin film preparation experiments were performed in a small two-chamber system, constructed from standard parts. The system consisted of a small load-lock and a main chamber. The main chamber was equipped with two effusion cells, pressure gauge and a sample manipulator. The sample carrier had a heating included, which could reach temperatures of up to 360 °C with 1.5 A (cp. Fig. 3.5). The chamber has been disassembled to reuse the components for the main MBE system and the test chamber.

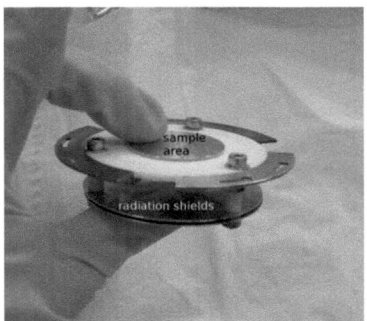
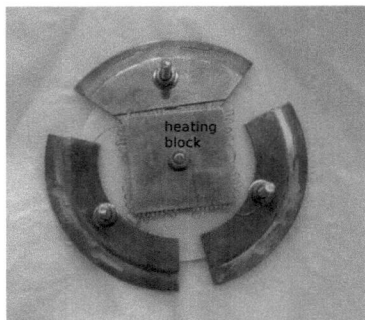

Figure 3.5: Side view and bottom view of the sample carrier

The heating block consists of a tungsten filament encased with ceramic tubes and embedded into a massive copper block. The heating block is directly attached to one copper segment and the sample area. The other two copper segments are electrically isolated from the heating block and provide the electrical connection for the heating if mounted on the sample manipulator.

3.6 Test chamber

This chamber is now a stand-alone mini MBE system, which was originally designed for leak testing of new components before integration into the MBE system. It has one effusion cell port, which is mounted to a linear travel and a sample holder, which houses microscope slides, and is also mounted on a second linear travel (Fig. 3.6). The chamber features a viewport mounted directly above the sample holder and therefore, using microscope slides, one can directly observe the deposition process. Furthermore, a series of samples can be prepared by shifting the microscope slide relative to the effusion cell with the linear travel. The sample-cell distance can be adjusted with the second linear travel, allowing for the study of evaporation characteristics in different distances.

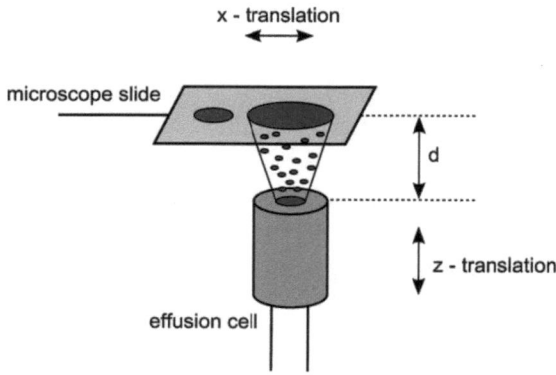

Figure 3.6: Schematic drawing of the test chamber
The microscope slide can be moved laterally and the effusion cell distance d is adjusted vertically, so that a series of experiments can be conducted onto one microscope slide.

Chapter 4

Effusion cells

4.1 Cell types and construction

Effusion cells are standard components in MBE for the evaporation of materials. An effusion cell usually consist of

- a crucible holding the source material to be evaporated,
- a heating mechanism - often realized by a heating coil (cp. Fig. 4.1) and
- radiation shields, concentrating the heat onto the crucible.

Since the field of organic electronics is gaining interest, commercially available effusion cells have been adapted for organic materials. Special requirements have to be fulfilled for organics, such as low temperatures - starting from about 120 °C up to about $500 - 600$ °C, small volumes (for small amounts of material) or good temperature homogeneity. Four basic types of effusion cells were developed, addressing different problems.

4.1.1 Ceramic tube cell

This first cell type was developed with special focus on temperature homogeneity. A tungsten wire is used as a resistive heater. The wire is wound onto a ceramic tube, which is supposed to

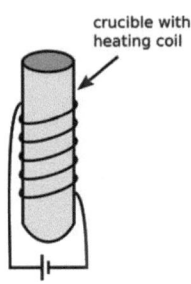

Figure 4.1: Sketch of the simplest form of an effusion cell

provide thermalization.

The crucible is a standard Al_2O_3 crucible of 8 mm diameter and 100 mm length. It is held by a metal cage, positioned at the center of the ceramic tube without contact to the walls, and therefore heated via radiation. Two metal cylinders are mounted surrounding the ceramic tube as radiation shields and are screwed to a base plate, simultaneously fixing the assembly. The parts can be seen on the photograph in Fig. 4.2.

The tungsten heating wire was formed to a coil, winding the wire onto a rod using a turning lathe. The coil was then transferred to the ceramic tube. The forming rod had a diameter of 13.5 mm and the windings were wound closely-packed onto the rod. The parameters for forming the coil are not very crucial, since the ceramic tube stabilizes the coil. The fit on the tube was tight and the individual windings could be arranged accurately with about 2.5 mm distance in between and a total of about 50 windings.

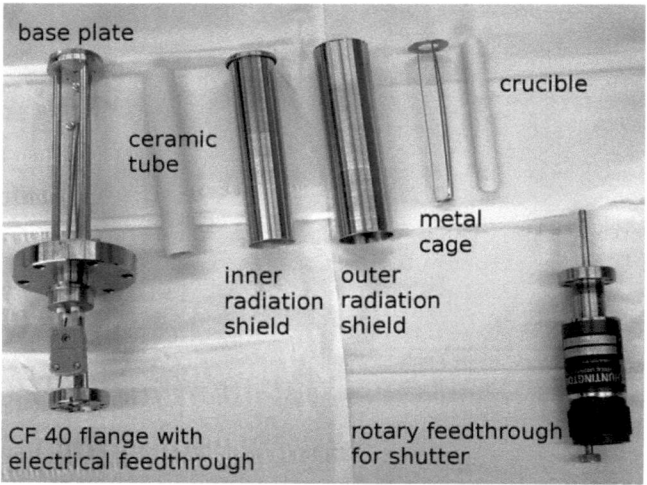

Figure 4.2: Components of the ceramic tube cell

On the very left the CF 40 flange, on which the cell is mounted, is shown. It has a four-fold electrical feedthrough (2× current and 2× thermocouple). On the attached CF 16 flange the rotary feedthrough for a cell shutter is mounted. The shutter itself is not depicted. On the vacuum side of the CF 40, three spacer and a base plate are already mounted. On top of this the ceramic tube is fixed by applying the inner radiation shield and finally by screwing the assembly together with the outer radiation shield. The heating coil, which has to be fit onto the ceramic tube, is not shown. Finally, the metal cage and the crucible are inserted trough a hole in the top of the outer radiation shield. The crucible is easily exchangeable.

A type-K thermocouple was spot-welded to the bottom of the metal cage for temperature measurement. With a current limitation to 4 A a maximum temperature of 900 °C was reached. For operation in this temperature range it is strongly recommended to use a cooling shroud.

The integration of the ceramic tube has the direct disadvantage that the cell has very slow response times, especially in the temperature range of 50 °C-150 °C. The problem can be

Figure 4.3: Schematic view of a free-filament cell

In the center is the crucible (yellow), held by a metal cage. Then there is the tungsten coil, held by 3 ceramics, which are not depicted here. The whole cell body is closed by two radiation shields. The thermocouple for temperature control is inserted through the central bore of the base plate and attached to the lowest point of the metal cage, right at the bottom of the crucible.

compensated upon heating by the use of a PID controller, but not upon cooling down.

4.1.2 Free-filament cell

To reduce response time a free-standing filament cell was developed. The principle setup is the same as for the ceramic tube cell, but the massive tube is exchanged with three thin ceramic rods in a triangular arrangement. A more rigid tungsten wire of thickness 0.38 mm was used, since the filament now has to support itself. A schematic view of an assembled cell is shown in Fig. 4.3.

The coil was formed on a rod of 13.5 mm diameter with the windings closely-packed, resulting in an inner diameter of 20 mm and a distance of $3-4$ mm between the windings.

The three thin ceramic tubes (outer diameter 2.5 mm, inner diameter 1.3 mm) fix the inner diameter of the heating coil. Otherwise the filament is supported by one ceramic, which guides the upper open end of the coil back down to the electrical feedthrough.

This cell shows significantly faster response times, but is much more difficult to assemble. A further disadvantage is that the top most windings of the heating coil are usually separated further than the rest due to the weight of the coils. This enhances one known problem of effusion cells, namely the inhomogeneous temperature distribution in the cell. The cell hosts a standard crucible of 8 mm diameter and 100 mm length.

One free-filament cell with cooling shroud was operated up to 1235 °C with a current limitation of 7 A. At this temperature the melting of the stainless steel metal cage resulted in the destruction of the filament. For continuous operation, temperatures above $1000 - 1100$ °C are therefore not recommended, unless the hot metal parts are manufactured from a high-melting point material, e.g. Tantalum.

4.1.3 Graphite cell

As an alternative to the filament cells, different graphite cells were developed. The basic idea is that a graphite crucible simultaneously acts as the resistive heater.

Therefore a crucible was manufactured from a graphite rod[1]. Electrical contacts have to be

---
[1] http://www.tryba-stockum.de, graphite quality TS 400-31

connected to the top and bottom of the crucible. The bottom contact is realized with a thread which allows for screwing the crucible into a metal base plate. The top contact is fastened with a metal clamp.

To achieve temperatures up to $400 - 500\,°C$ with a current below 15 A a relatively thin wall thickness is necessary, because of the rather low resistivity $\rho = 17 \cdot 10^{-2}\,\Omega cm$ of graphite. Three designs with different outer diameter d_o and inner diameter d_i were studied.

1. $d_o = 15$ mm, $d_i = 10$ mm: With 7 A a temperature of 190 °C was reached.

2. $d_o = 6$ mm, $d_i = 5$ mm: With 7 A a temperature 300 °C was reached, but the crucible was too fragile for permanent operation.

3. $d_o = 12$ mm, $d_i = 10$ mm, with $d_i = 11$ mm at the top of the cell: With 7 A a temperature of 200 °C was reached, but the thinner part of the crucible was too fragile and broke off.

The final design is a variation of No. 3, using a constant inner diameter of $d_i = 10$ mm from bottom to top. This is still a rather fragile crucible and one has to be very careful to avoid strain, especially when attaching the top contact. While testing the cell with copper-phthalocyanine

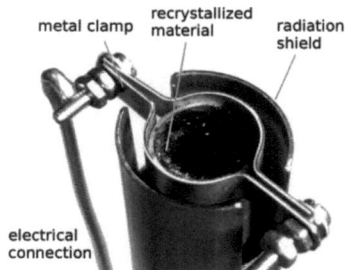

Figure 4.4: Photograph of a graphite cell

The graphite cell body is shown with the attached metal clamp. The electrical connection is a massive copper wire coated with silver. It is on this picture only loosely connected and is usually fixed between the two nuts on the left side of the metal clamp. A radiation shield is mounted, with openings for the metal clamp. CuPc crystallites are growing at the top of the crucible and reduce the diameter of the cell. Without additional heating at the top of the cell, the cell closes after a few deposition processes.

(CuPc), material recrystallized at the upper edge of the cell and caused the cell orifice to close (cp. Fig. 4.4). This is a known problem for many materials stemming from the fact that the open end of the cell is usually colder than the very bottom of the crucible. For this effusion cell type this is presumably caused by the metal clamp at the top of the cell. The clamp has a certain width and the current flows directly through the clamp instead of flowing through the top of the cell. The attempt to compensate this problem with an region of smaller cross section at the top failed due to mechanical instability.

A solution for the problem is an extra tungsten wire, which is attached as an additional heater for the cell top. It is electrically connected in series with the graphite body. This way the recrystallization at the top of the cell was drastically reduced.

4.1.4 Mini cell

This effusion cell was designed by Prof. Dr. Michael Huth for small amounts of source material. It is especially suited for the study of new materials, for which sublimation point and temperature stability are unknown. The combination of a small crucible and a massive body provides very good temperature homogeneity.

The cell hosts a crucible of 13 mm diameter and 13 mm length. The cell body is massive aluminum with one bore for the crucible, one bore for a thermocouple and a milled-into structure to host a thermo coax cable. Aluminum was chosen for good heat conduction. The assembly is fixed with an aluminum casing, which also acts as a radiation shield. Additionally an exchangeable lid covers the top of the cell and allows for different orifices.

Experimentally it turned out that the temperatures for mini cell and free-filament cell are usually not directly comparable. It was found, that the same materials sublimed at nominally 50 °C higher temperatures in the mini cell as in a free-filament or the ceramic tube cell. Different free-filament cells and the ceramic tube cell were found to be comparable in this respect. This demonstrates that sublimation temperatures measured with effusion cells should always be examined carefully, since the measured temperature strongly depends on the precise position of the thermocouple and other factors such as the crucible size and material and the filling level.

4.2 Temperature measurements

To compare the homogeneity of the different cells and to obtain information for the Monte Carlo simulation presented in the next chapter height-dependent temperature measurements (T-measurements) were performed.

Experiment 1

An aluminum oxide crucible was equipped with three type-K thermocouples in different positions inside the crucible. Each thermocouple was attached to a metal ring, which fitted the crucible diameter. They were positioned at the bottom of the crucible, the top of the crucible and the middle position in-between. The crucible was inserted into a free-filament cell. Data for different temperatures are shown in Fig. 4.5. The reference temperature measured with the thermocouple attached to the metal cage (cp. section 4.1.1) is not included, because the thermocouple was damaged and not working properly during the experiment. The temperatures measured at middle and bottom position differ less than 5% and the relative difference is rather stable. In contrast, the temperature difference between top and bottom of the cell increases significantly with temperature. At the highest temperatures studied the relative difference reaches around 20%.

Experiment 2

To refine the measurement and obtain a temperature profile instead of single data points the experimental setup was modified. The effusion cell under investigation was mounted onto a linear travel and a type-K thermocouple was inserted into the crucible. During the experiment the crucible was empty and the thermocouple was pressed to the inner wall of the crucible with

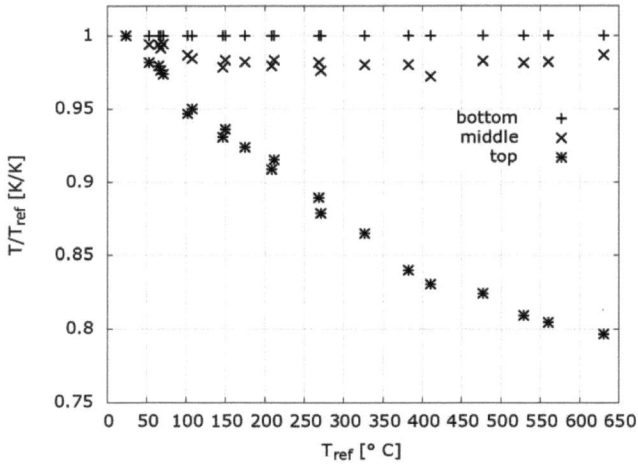

Figure 4.5: T-measurement of a free-filament cell with Al$_2$O$_3$ crucible
The temperature at three different positions in the crucible (top, middle, bottom) was measured. Depicted is the temperature of the middle/top position relative to the temperature at the bottom of the cell. The temperature gradient between top and bottom increases with increasing cell temperature. For the highest temperature investigated $T_{\text{ref}} = 631$ °C the temperature at the top of the cell reaches only about 80 % of the bottom temperature, where the sublimation takes place.

a metal spring. With the linear travel the depth of the thermocouple inside the crucible could be varied and the temperature was measured at different positions inside the cell.

One problem of this setup is that the thermocouple also acts as a heat conductor. Therefore the measured data should be regarded as a lower limit for the real temperature. The same holds true, if the thermocouple looses close contact to the wall during the measurement.

Free-filament cell with graphite crucible A graphite crucible with the dimensions of a standard aluminum oxide crucible with length 100 mm, outer diameter 10 mm and inner diameter 8 mm was manufactured. Additionally the top of the crucible was thinned to an outer diameter of 9 mm for a length of about 10 mm. This prevents contact between the crucible and the radiation shields at the top of the effusion cell. The question was, whether this reduces heat conduction at the top of the cell and therefore improves the temperature homogeneity.

The experiments show that the use of the graphite crucible further increases the temperature gradient from bottom to top compared to the standard aluminum oxide crucible (cp. Fig. 4.6). For the three bottom temperatures 270 °C, 350 °C and 535 °C the temperature profile is very similar and the temperature at the very top of the crucible is about 60 % to 70 % of the temperature at the bottom. This is worse than the gradient for the aluminum oxide crucible, especially at low temperatures. Whether this behavior stems from the new design, purely from the material properties or a combination of both was not investigated further.

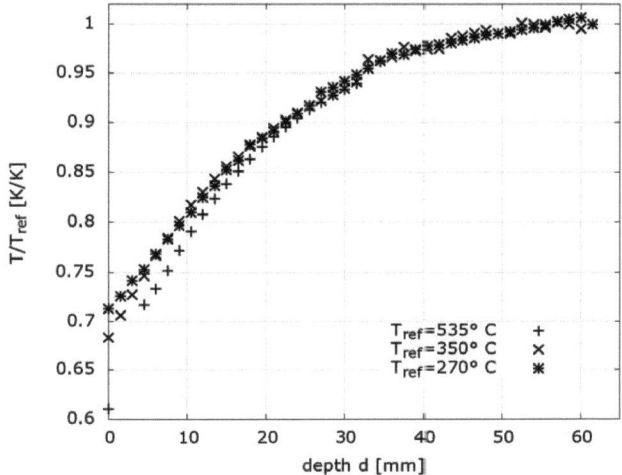

Figure 4.6: T-measurement of a free-filament cell with graphite crucible
The temperature profiles for the graphite crucible at three different temperatures are shown. $d = 0$ mm corresponds to the top of the crucible.

Ceramic tube cell with aluminum oxide crucible For the ceramic tube cell with aluminum oxide crucible four temperature profiles were measured (cp. Fig. 4.7). Here the reference temperature is the temperature measured by the build-in thermocouple of the cell. The cell was operated at constant temperature controlled by an Eurotherm PID controller. Two tendencies can be seen:

1. The higher T_{ref}, the better the agreement of T_{ref} and $T_{d=80\text{ mm}}$.

2. The higher T_{ref}, the stronger the gradient between $T_{d=80\text{ mm}}$ and $T_{d=0\text{ mm}}$.

Free-filament cell and ceramic tube cell both show rather strong temperature gradients along the crucible and the expected thermalization of the ceramic tube could be disproved by the experiments. To obtain a homogeneous temperature the loss at the open end of the cell has to be compensated by a stronger heating at the top of the effusion cell.

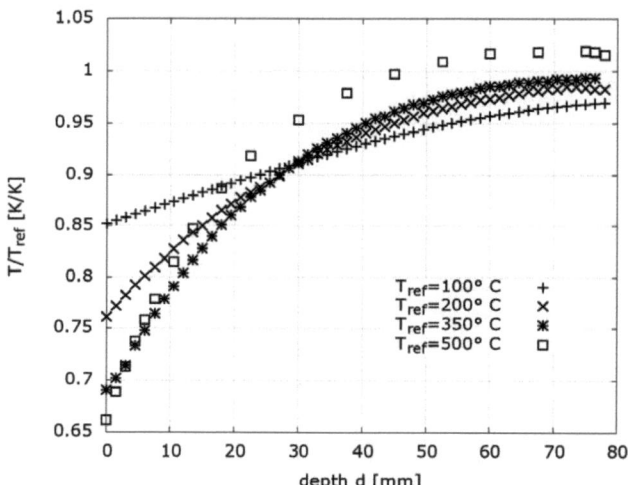

Figure 4.7: T-measurement of the ceramic tube cell with Al_2O_3 crucible
The position $d = 0$ mm corresponds to the top of the cell, $d = 80$ mm corresponds to the deepest point reachable with the experimental setup. T_{ref} is the reference temperature measured with the build-in thermocouple of the cell.

Chapter 5

Evaporation Characteristics

5.1 Introduction

In MBE and OMBD a variety of different evaporation cells are used. The two most common types are so-called effusion cells and Knudsen cells. Both usually host a cylindrical crucible with large aspect ratio. The Knudsen cell differs from the effusion cell by having a lid with a small orifice to close the crucible. Since the term *effusion cell* is often used synonymously for both Knudsen and effusion cells, in the following *open cell* is used for a simple cylindrical crucible and *Knudsen cell* for a crucible with a lid with a small orifice. A sketch of both types can be seen in Fig. 5.1. The decision to use one or the other type of cell is mostly based on personal experience and preference rather than physical considerations.

To overcome this practice, a Monte Carlo program for the simulation of evaporation characteristics of different cell types was developed. On basis of the simulation results, the design of a cell shall be determined according to the desired evaporation characteristics and, accordingly, the deposition profile at the sample. To achieve this goal an understanding of the relevant processes is necessary, which include evaporation, migration and collisions. After basic functional tests of the program, several cell geometries were studied to optimize the geometry for materials with low evaporation rates, which are thermally already close to decomposition. To validate the results experiments were performed, accordingly.

Finally, the program is not limited to the field of molecular beam epitaxy, but can be extended to any field dealing with evaporative particle beams and a constraining geometry. A second field of application, electron beam induced deposition, is presented in Appendix B.

5.2 Theory

Martin Knudsen was one of the pioneers, regarding the theory of the flow of gas through tubes. He has derived the *cosine law* for the reflection of atoms from tube walls [47, 48], the evaporation from atoms from a metal source [49] and has introduced the geometry, which is today known as the Knudsen cell [48].

The Knudsen number $K = \frac{\lambda}{L}$ characterizes different flow regimes, where λ is the mean free

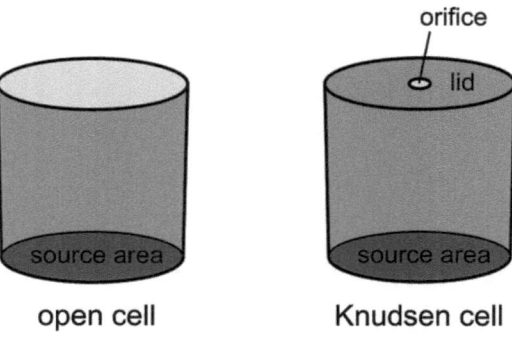

Figure 5.1: Sketch of an open cell and a Knudsen cell
The open cell is a cylindrical crucible with the source material at the bottom of the cylinder. The Knudsen cell is identical with an additional lid at the top of the cylinder, which has a small orifice.

path of the gas particles in the cell and L is a characteristic length scale. In the case of open cells/Knudsen cells the characteristic length is usually either diameter or length of the cell.

- $K \gg 1$: molecular flow regime. Intermolecular collision can be neglected and only particle-wall collisions have to be taken into account. Open cells are assumed to operate in this regime.

- $K \ll 1$: viscous flow regime. λ is small compared to L and hence the particle density inside the cell is so high that the gas flow approaches the conditions in a liquid.

- $K \approx 1$: transition flow regime. This regime marks the transition between molecular flow and viscous flow and is least well understood.

For an ideal Knudsen cell, defined by an infinitely small orifice with negligible thickness, the evaporation characteristic can be determined analytically. Under the assumption of thermal equilibrium inside the cell the *cosine law* is obtained. Particles in the enclosed volume obey the Maxwell-Boltzmann velocity distribution

$$F(v^2) = \sqrt{\frac{2}{\pi}} \left(\frac{m}{2\pi k_B T}\right)^{3/2} v^2 \exp\left(\frac{-mv^2}{2k_B T}\right), \tag{5.1}$$

where v is the velocity of the particles, m the mass, k_B Boltzmann's constant and T the temperature. The number of particles effusing into a solid angle $d\Omega$ in the time interval dt is

$$d^4 N = N_0 \cdot \frac{\cos(\vartheta) \cdot v \cdot dt \cdot dA}{V} \cdot F(v^2) dv \cdot \frac{d\Omega}{4\pi}, \tag{5.2}$$

where N_0 is the total number of particles in a cell of volume V and dA is the size of the orifice. The number of particles received on an area dS in the distance r from the source is, with

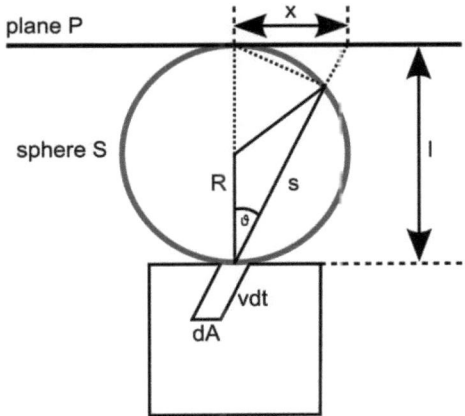

Figure 5.2: Cross section of a Knudsen cell
A Knudsen cell with orifice dA is shown. A flat plane is situated in the distance l from the orifice. Additionally, a sphere with the diameter $2R = l$ is depicted. Particles from the volume $dA \cdot v \cdot dt$ inside the Knudsen cell can leave through the orifice into the solid angle defined by ϑ. Particles hit the plane at a distance x from the center of the plane, which is directly above the orifice of the cell.

$dS = r^2 d\Omega$,

$$d^4 N = N_0 \cdot \frac{\cos(\vartheta) \cdot v \cdot dt \cdot dA}{V} \cdot F(v^2) dv \cdot \frac{dS}{4\pi r^2}. \tag{5.3}$$

One can therefore define the flux per unit area and unit time at distance r and angle ϑ

$$\Gamma(r, \vartheta) = \Gamma_0 \frac{\cos(\vartheta)}{\pi r^2}, \tag{5.4}$$

with Γ_0 the flux at $\vartheta = 0$. This is known as the cosine law, or Knudsen law, and is analogous to Lambert's law in optics.

Knudsen has shown that this evaporation characteristic leads to a homogeneously covered sphere, where the orifice lies on the surface of the sphere, and has used this relationship to verify the validity of the cosine law experimentally [50]. This is derived in the following paragraph. Refer to Fig. 5.2 for geometry and nomenclature.

Geometrically one obtains that the flux through a point on the surface of a sphere of radius R at distance $s(\vartheta) = l \cos(\vartheta)$ from the orifice in terms of ϑ is

$$\Gamma(s(\vartheta), \vartheta) = \Gamma(l\cos(\vartheta), \vartheta) = \Gamma_0 \frac{1}{\pi l^2 \cos^2(\vartheta)} \cdot \cos(\vartheta) = \Gamma_0 \frac{1}{\pi l^2 \cos(\vartheta)}. \tag{5.5}$$

Defining a unit area dA_0 orthogonal to $s(\vartheta)$ one has to take into account that the projection dA_S of this area onto the sphere varies in size with ϑ as

$$\frac{dA_0}{dA_S} = \cos(\vartheta). \tag{5.6}$$

Combining equation (5.5) with (5.6) one obtains that the flux on the surface of the sphere

$$\Gamma_S(s,\vartheta) = \Gamma(s,\vartheta) \cdot \frac{dA_0}{dA_S} = \Gamma_0 \frac{1}{\pi l^2 \cos(\vartheta)} \cdot \cos(\vartheta) \tag{5.7}$$

is constant according to Knudsen's theory and experiments.

For practical issues an interesting geometry is a plane P in the distance $l = x \tan(\vartheta)$ from the orifice. A point on the plane at distance x from the center of the plane has the distance $L = \frac{l}{\cos(\vartheta)}$ from the source. Therefore the flux on the plane at the distance L and angle ϑ is given by

$$\Gamma(L,\vartheta) = \Gamma(L,0) \cdot \cos(\vartheta) = \Gamma\left(\frac{l}{\cos(\vartheta)}, 0\right) \cos(\vartheta) = \Gamma(l,0) \cdot \cos^3(\vartheta) \tag{5.8}$$

this flux hits the area dA_p on the plane where

$$\frac{dA_0}{dA_P} = \cos(\vartheta) \tag{5.9}$$

and therefore the flux on the plane is

$$\Gamma_P(L,\vartheta) = \Gamma(L,\vartheta) \frac{dA_0}{dA_P} = \Gamma(L,\vartheta) \cos(\vartheta) = \Gamma(l,0) \cos^4(\vartheta). \tag{5.10}$$

This $\cos^4$-distribution serves as a reference for the simulations. The flux distribution is here equivalent to a particle density and in the experiments with the film thickness at a given point.

5.3 Monte Carlo Program

5.3.1 Introduction

The aim of the program is to simulate evaporation characteristics of open cells, Knudsen cells and new cell types. It is intended to simulate different cell geometries instead of performing time consuming and expensive experiments to evaluate the evaporation characteristics.

5.3.2 Monte Carlo method

A Monte Carlo simulation is a simulation of a stochastic process using a sequence of random numbers [51]. The random numbers are usually generated by a pseudo-random number generator. Conventionally, random numbers in the interval $[0, 1[$ are generated and the quality of the pseudo-random number generator can be determined by

- the uniformity of the generated numbers over the interval and

- non-correlation between consecutive numbers.

Non-uniform random numbers, e.g. following a Gaussian distribution, can usually be derived from a uniform random number generator using a mathematical transformation.

All simulation results in Monte Carlo rely heavily on the quality of the random number generator. In this work the Mersenne twister algorithm was used. It is known for a very long period and high order of equidistribution [52] and is therefore highly approved.

It should be mentioned that the simulation works with dimensionless variables, but it is assumed that all lengths can directly be converted to mm, which is a reasonable order of magnitude for the comparison to experiments.

5.3.3 Geometry

Concerning the experimental setup, the focus lies on fully cylindric geometries, where the ϕ-dependence could in principle be omitted. Nevertheless, the simulation contains the full cylinder geometry, which allows for checking, whether the cylinder symmetry is conserved during the simulation. Additionally, this enables us to simulate substrates tilted relative to the symmetry axis of the cell and opens the possibility to include non-cylindric elements in the future.

5.3.4 Program structure

The main processes relevant for the problem are evaporation of particles from the source material, their migration and collisions. The simulation is two dimensional in the sense that only the crucible surface is modeled, but not the volume. The crucible is represented by a 2D surface grid, an example for the cylinder wall is shown in Fig. 5.3. One should note, that the geometry and the models of the physical processes are separated, so that each can be altered independently, which leaves potential for improvement and further development of the program. This property has already been exploited for the test of different collision models.

5.3.5 Basic elements

The program is written object oriented. The main objects are

- **regions:** The crucible geometry is described by a number of regions and the symmetry axis is defined as the z-axis. A region consists of a grid of surface elements each of which is characterized by size A, temperature T and coverage N. The coverage is an integer number counting the number of particles on the surface element. The region *type* enables to distinguish between different geometries. At the moment incorporated types are

    - **start:** This is a circular, flat disc, describing the ground of the crucible, containing the source material. The relevant geometric parameter is the radius. It is an infinite source of particles, with constant coverage. The surface normal is pointing along the z-direction.

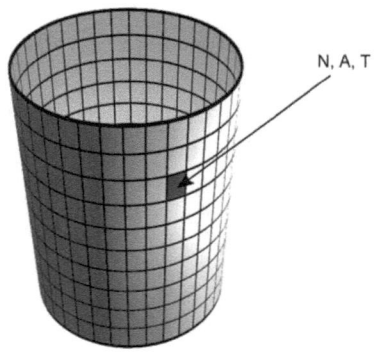

Figure 5.3: Cylinder grid
The cylindrical walls of a crucible are represented in a two dimensional array, where the respective geometric position is stored, as well as the additional attributes coverage N, area A and temperature T.

- **bottom:** A circular, flat disc or ring, geometrically similar to a start region, but without being an infinite source and characterized by an outer radius and inner radius. The surface normal is pointing along the z-direction.

- **side:** A cylinder wall, characterized by radius and length. Here the axis of the cylinder is parallel to the z-direction.

- **top:** Equivalent to *bottom*, with the difference that the surface normal is pointing along the $-z$-direction. This region represents the lid.

- **particles:** A particle is characterized by a position vector $\vec{p}$, which determines the actual position of the particle, and a direction vector $\vec{d}$, which describes the direction of movement, equivalent to a velocity vector. The position of a particle is usually located on a surface element and the direction is first determined with respect to the surface normal of this surface element. Coordinates are then transformed to a global coordinate system. In this global system the intersection of the *particle direction* with the crucible geometry is calculated. Furthermore the particle carries information about the number of scattering processes it has suffered and the position/probability of a subsequent scattering process.

- **planes:** A flat square plane, representing the substrate. The surface normal may be pointing along the z-direction or is tilted with respect to z with an angle α. The plane is divided into equal quadratic area elements and the center of the plane, $(x = 0, y = 0)$, is fixed to the intersection point with the symmetry axis of the crucible. The coverage of the plane is stored as a two dimensional histogram, and independently as an one dimensional radius-dependent histogram. Both data sets are calculated separately, so that the binning of one data set does not influence the other.

The regions composing the whole crucible are organized in a linked list of pointers, which is an appropriate description for using an iterative way of calculating the intersection of a particle direction with the different regions. The advantage is that the program is capable to deal with a variable number of regions, allowing for complex crucible geometries.

To implement new region types the coordinate transformation from and to the global coordinate system and rules for the calculation of the intersection of a straight line with the respective surface have to be defined.

5.3.6 Program flow

In the beginning the geometry is initialized. Then the program iterates the following steps:

1. A surface element is randomly chosen and - if occupied - a particle may move through evaporation or migration.

2. A second random number determines whether evaporation or migration occurs. By simulating the chosen process a final surface element is identified, which may be part of the crucible or part of the receiving substrate area. An evaporation process may hereby be accompanied by a number of scattering events, which is treated in detail in section 5.3.9.

3. The particle is removed from the start element and added to the final surface element.

4. A new surface element is chosen (cp. Fig. 5.4).

Step number 4 distinguishes this *single-step model* from the works of [53, 54, 55, 56], where one particle is observed until it leaves the cell, which might include multiple evaporation and wall absorption processes; this model is in the following called *trajectory model*. For homogeneous temperatures and constant evaporation/migration energies the two models are equal and differ only in the chronology of the calculation steps.

As soon as temperature gradients are considered along the cell, the two approaches are more distinguishable. In the *single step model* the temperature of a chosen area element directly determines the evaporation probability for the occupying particles through a Boltzmann factor. Accumulation of particles on *cold* surface elements occurs therefore naturally. That this effect can not be neglected has also been demonstrated in section 4.1.3, where recrystallization at the top of an effusion cell was shown. In the *single-step model* the crucible wall strives to a stable coverage and provides a stability criterion for the simulation. In contrast to this, in the trajectory model generally all particles leave the cell, so there is no coverage of the crucible. Temperature effects are sometimes included by a *sticking* factor [57], oppressing evaporation from certain surface elements.

5.3.7 Evaporation

The evaporation probability is determined by a Boltzmann factor $e^{\frac{-E_{vap}}{k_B T}}$, where the directional dependence is determined by the cosine law. This is achieved by choosing random numbers r_1

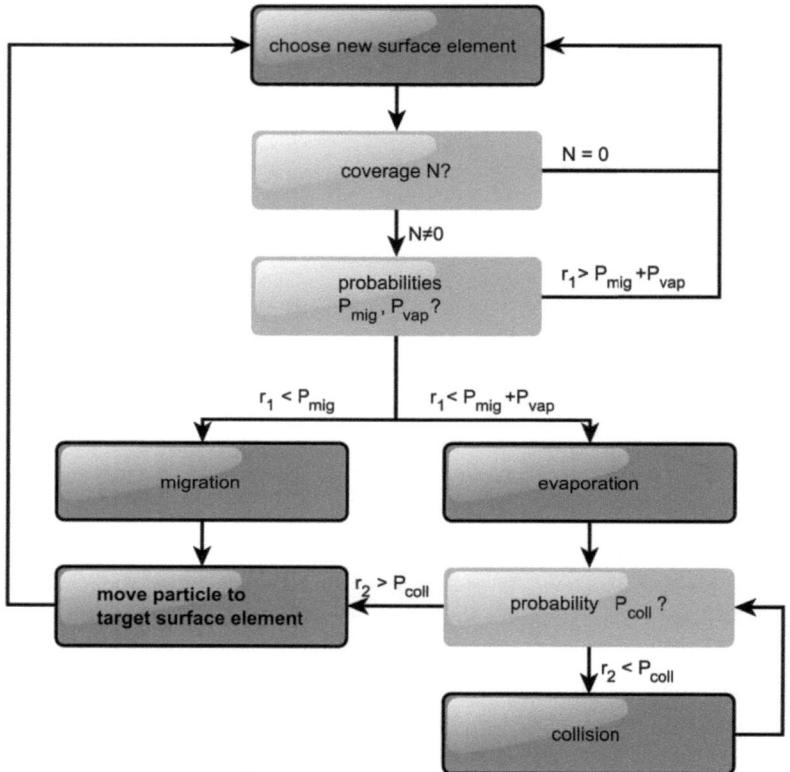

Figure 5.4: Program structure
A surface element is randomly chosen. If it is occupied, the probabilities for evaporation, P_{vap}, and migration, P_{mig}, are calculated with respect to area A, temperature T and coverage N of the surface element. A random number r_1 determines which process occurs. For migration the particle is moved to one of the neighboring surface elements. For evaporation the direction is randomly chosen (according to the cosine law) and the intersection with the crucible geometry is determined. Optionally, collisions may occur during the evaporation process. Finally, a particle is moved to the target surface element.

and r_2 with $\phi = 2\pi \cdot r_1$ and ϑ according to $f(\vartheta) = \sin(2\vartheta)$. The direction vector is then

$$\vec{d} = \begin{pmatrix} \sin(\vartheta)\cos(\phi) \\ \sin(\vartheta)\sin(\phi) \\ \cos(\vartheta) \end{pmatrix}$$

and the piercing point of the particle's trajectory is determined by calculating the intersection of the straight line $\vec{g} = \vec{p} + m \cdot \vec{d}$ with the geometry of the crucible/plane. The particle is then removed from the starting surface element and added to the final surface element determined by the intersection.

5.3.8 Migration

The probability for migration is also determined by a Boltzmann factor $e^{\frac{-E_{mig}}{k_B T}}$. A particle is allowed to move to any of the neighboring surface elements with equal probability. Neighboring surface elements are currently defined as surface elements sharing an edge with the start element. The particle can migrate from one region to the next, taking into account the correct boundary conditions between the regions. A different possibility is to define a certain migration radius and allow for movement inside this circle. On the substrate - represented by planes - migration is not considered. Physically, migration on the substrate influences the thin film growth, but the focus here is on the evaporation characteristic of the effusion cells and not on growth on the substrate. A homogeneous evaporation rate over the area of the substrate promotes uniform growth on the substrate independent of the physical processes happening on the substrate.

5.3.9 Collisions

Although the approximation of molecular flow is suitable for the simulation of open cells, one should keep in mind that the cosine evaporation characteristic of a Knudsen cell stems from the fact that there is thermalization and accordingly collisions in the enclosed volume of the cell. The expectation is hence that the simulation of a Knudsen cell is only correct if one explicitly includes collisions. Additionally, collisions allow for exploring the capability of the program to simulate not only the molecular flow regime, but also possibly the transition regime. There are several problems connected with collisions. First of all, physically a particle scatters with a second particle. Secondly, these collisions occur in the volume of the cell. In the program only one particle is observed at a time and it was originally not intended that this particle stops/stays in the volume of the cell, therefore the geometry is limited to the surface area of the crucible. The problems were solved in the following way: since a collision occurs *only* in the volume, and not on the surface, collisions can be coupled to the evaporation process. An effective model has to be found, providing a collision point or collision length, determining the position, where a particle scatters during the evaporation process. The scattering has to be iterated, until the particle hits either the crucible walls or leaves the cell. This means an evaporation process may now be accompanied by multiple scattering events. Two collision models were implemented and are treated in detail in section 5.5.

5.3.10 Simulation Output

Various data sets are generated within one simulation. The most relevant data files are shortly explained for a better understanding of the graphs shown in the following. All files are simple ASCII data.

configuration file All relevant data about the start and end parameters of the simulation is stored, such as geometry of the cell, number of iteration steps, number of evaporated/migrated/scattered particles, temperature and energies, as well as the seed for the random number generator. The data stored here is sufficient to rerun the simulation with the exact same parameters.

plane grid data The coverage of a plane is stored as a two dimensional histogram. This means the central position of every area element on the plane is stored together with the respective number of particles, providing an (x, y, N) data set. This data is used to show a three dimensional thickness distribution on the plane.

plane radius data The radius data contains data sets of the form $(r, N(r))$, where $r = \sqrt{x^2 + y^2}$, with (x, y) coordinates of the hitting point of a particle on the plane. This is calculated independently from the grid data. Every particle is sorted into a histogram according to r, with the bins corresponding to rings on the plane. The number of particles is corrected by the area of the respective ring, to obtain a radius-dependent coverage. This approach provides a better overall statistics compared to a cross section of the plane, because all particles are accounted for, but because of the division through the area, the relative error for small values of r is large. If the simulated problem is not cylinder symmetric, e.g. by simulating a tilted substrate, this data corresponds to the thickness distribution when a sample rotation is employed.

region grid data The coverage of the different regions of the cell is stored. This is useful to observe migration or the influence of temperature. As default the first *side* region is stored in this way. This cylindrical region can then be displayed as a plane, where the x coordinate of the plane codes the height z and the y coordinate codes the angular position.

5.3.11 Simulation Results

Point source

As a first test of the correct implementation of the cosine law a cell with a single point as evaporation source was simulated - in the following abbreviated as point source. An isotropic emitter was implemented for comparison. To visualize the results the radius-dependent coverage

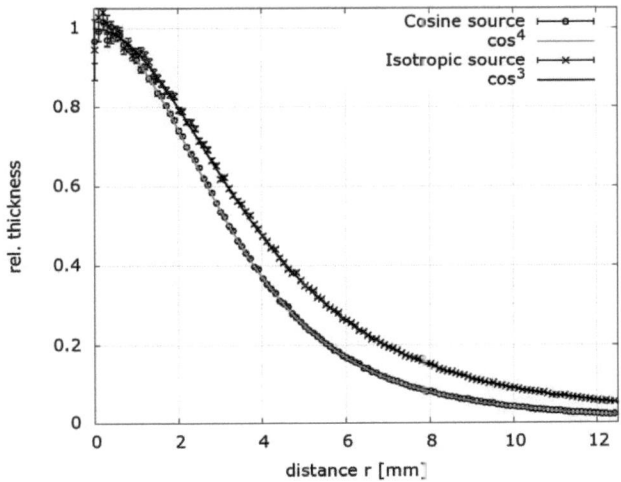

Figure 5.5: Point source vs. isotropic source
The *plane radius data*, meaning rel. thickness versus distance r from the center of the plane, is shown for the simulation of an isotropic source and a point source. For comparison the theoretical expected $\cos^3$ and $\cos^4$ distributions are also depicted. The simulation shows perfect agreement with theory.

of the substrate plane is shown in Fig. 5.5, where the x-axis depicts the distance from the center of the plane and the y-axis shows the relative particle coverage or, equivalently, the relative film thickness. The source-substrate distance is $l = 6$ mm, which is an arbitrary choice at this point, but is a suitable distance for the comparison to the experiments later. Both simulations show the expected behavior, $\cos^4$ for the point source and $\cos^3$ for the isotropic source, respectively.

Open cell

The next step was to simulate an open cell. The assumption is that an open cell operates in the molecular flow regime and particle collisions can therefore be neglected. An important measure for an open cell is the aspect ratio L/d of length and diameter of the crucible. A sketch of the crucible geometry is shown in Fig. 5.6. It has been shown by, e.g., Stickney et al. [58] that open cells exhibit a focusing of the molecular beam, which is more pronounced for larger aspect ratios. This behavior is nicely reproduced by the simulation as shown in Fig. 5.7. The effect of collisions on the simulation of an open cell is treated in section 5.5.

Knudsen cell

The advantages of a Knudsen cell are two-fold. If the sample is located in great distance from the source, one obtains a rather homogeneous growth over a relatively large area. The distance has to be so large that the flat sample is a good approximation of the homogeneously covered sphere (cp. section 5.2). The disadvantage directly connected to this is that a large portion

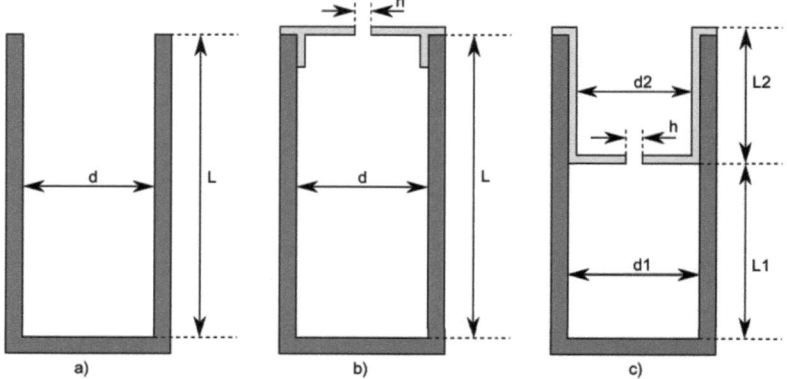

Figure 5.6: Cross sections of the studied crucible geometries
a) open cell, b) Knudsen cell, c) 2-stage cell
All cell types are based on a cylindrical body with an inner diameter d and an overall length L. For a Knudsen and a 2-stage cell a lid with an orifice (hole) of diameter h is present. For the two-stage cell the total length is divided into two segments of length $L1$ and $L2$, which differ in diameter. The diameter $d1$ corresponds to d for open and Knudsen cell. The lid thickness is assumed to be 0.

of the evaporated material is lost to the environment instead of contributing to the thin film growth on the sample.

The second advantage is that for materials close to decomposition, the Knudsen cell geometry seems to be more appropriate. We have observed higher evaporation rates with Knudsen cells compared to open cells. Whether this stems from the higher vapor pressure inside the cell or is a result of a better temperature homogeneity in the cell is not clear up to now. For the simulation the problem is that the cosine law relies on the thermal equilibrium in the cell, which means that collisions have to be taken into account. To identify the requirements for a suitable collision treatment two 3D simulations where performed:

a) In a cylindrical cell of Knudsen geometry (parameters $L = 8$ mm, $d = 8$ mm, $h = 1$ mm) particles are generated homogeneously distributed over the whole volume. According to an isotropic distribution a direction vector is assigned to each particle and it is evaluated, whether the particle can leave the cell through the orifice. Particles leaving the orifice are added to the substrate plane as usually, the other particles are discarded.

b) Now, the role of the Maxwell-Boltzmann velocity distribution is investigated. Additional to the (normalized) isotropic direction vector, the particles obtain an absolute velocity v, according to the Maxwell-Boltzmann distribution. For all particles leaving through the orifice the length of the path l is determined and compared to the velocity. Defining an arbitrary time t, only particles with $v \cdot t > l$ leave the cell, the others are discarded. This favors particles closer to the orifice and particles with higher velocities.

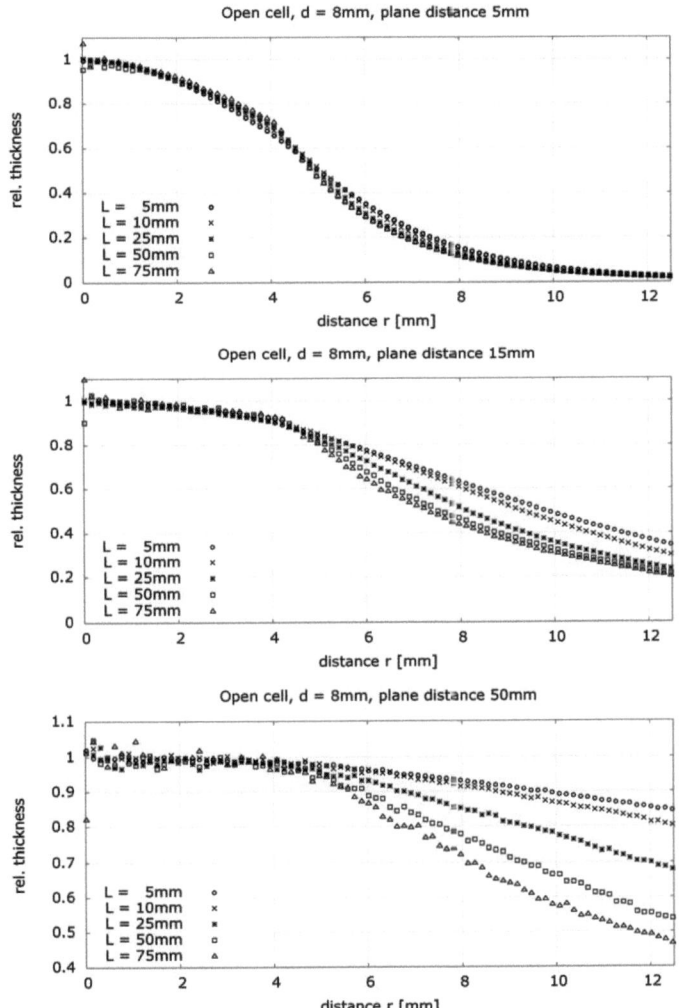

Figure 5.7: Simulation of an open cell
An open cell with diameter $d = 8$ mm and lengths from $L = 5$ mm to $L = 75$ mm was simulated. From top to bottom the results at a plane distance of $l = 5$ mm, $l = 15$ mm and $l = 50$ mm are shown. The longer the cell length, the stronger the focusing effect is visible. The effect is more distinct in a greater distance from the source. Since experimental distances range up to 250 mm, the open cell has clear advantages with respect to efficient source material usage.

The resulting distributions on the substrate are shown in Figure 5.8. The second simulation is in excellent agreement with the $\cos^4$-law expected for a Knudsen cell. This proves that the cosine law is not only governed by geometry, but also by explicit use of the Maxwell-Boltzmann velocity distribution. Inversely one can conclude that a collision model has to ensure two basic aspects to be reasonable: First, the collisions have to distribute the particle positions over the whole volume of the cell and, second, a distance-dependent probability for leaving the cell has to be provided. Here particle position means the last collision point, before the particle leaves the cell through the orifice.

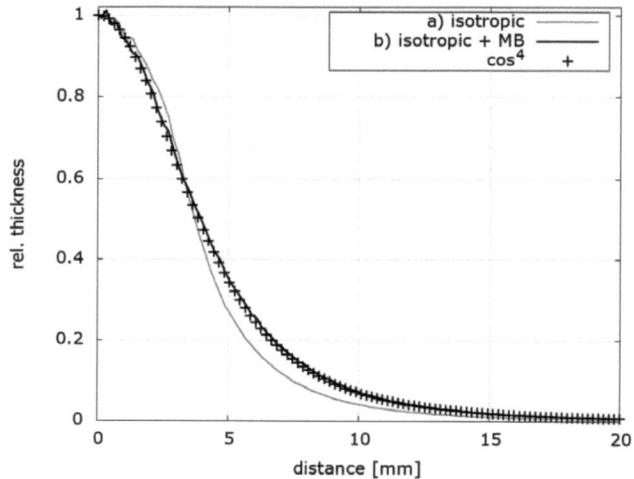

Figure 5.8: 3D simulation of the cosine law for a Knudsen cell
A Knudsen cell with $L = 8$ mm, $d = 8$ mm, $h = 1$ mm is simulated, with particles starting homogeneously distributed in the volume of the cell. Particles leaving the cell through the orifice are collected on a plane in 6 mm distance, the other particles are discarded. The graph shows the radial thickness distribution on the plane with respect to the point directly above the center orifice.

a) The pure geometric effect on the distribution is demonstrated.

b) Here additionally a velocity is assigned to each particle, according to the Maxwell-Boltzmann distribution, and only particles with sufficiently high velocity compared to their distance to the orifice are allowed to escape.

As a reference the theoretical result for the cosine law ($\cos^4$) is displayed. b) is in excellent agreement with the theoretical distribution.

2-Stage cell

To combine the advantages of open cell and Knudsen cell a new cell design was studied, which is in the following referred to as *2-stage cell*. The idea is to have a closed Knudsen-like crucible as a first stage of the cell, followed by a second stage, resembling an open cell. In the lower part one has the advantage of the higher vapor pressure and temperature homogeneity, whereas the upper part provides the collimation known from an open cell and hereby improves the yield of source material. A sketch of the geometry with the division into the respective regions is shown on the left hand side of in Fig. 5.9. Additionally, a simplified geometric model for the simulation is shown on the right hand side, where only the upper part of the 2-stage cell is simulated and the lower Knudsen-part of the cell is replaced by a *start* area in the size of the Knudsen orifice. The simplified model has two advantages. On the one hand simulation time is reduced, due to the smaller surface/volume to be simulated and, on the other hand, the simulation is independent of the collision model, since collisions can be neglected in the upper part of the cell.

A similar "crucible in crucible" geometry has been proposed by [59] to improve the uniformity of the deposition.

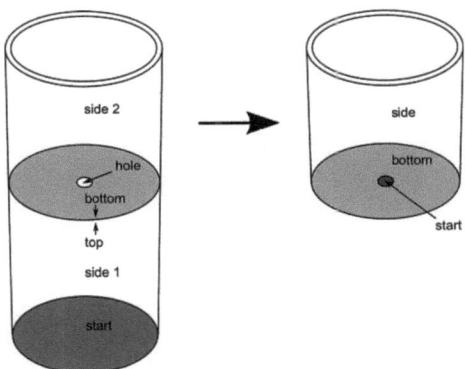

Figure 5.9: Geometric model of a 2-stage cell
The left sketch shows the complete model, with a full Knudsen cell as a lower part, consisting of *start* area, *side 1* and *top*, followed by a second wall segment *side 2* and a *bottom* area, which connects the two parts. On the right side there is a simplified model, where Knudsen behavior of the lower part is assumed. It is therefore replaced by a *start* area in the size of the hole of the original Knudsen part of the cell.

Comparison of the three different cell types

A simulation of each of the three cell types was performed to compare the evaporation characteristics and test the new 2-stage geometry. A comparison of the geometries is shown in Fig. 5.6. To reflect the experimental circumstances the total length of all cells is $L = 75$ mm, assuming a 100 mm standard crucible with typical filling level of source material, which can be modified with different apertures. The diameter of all cells is $d = 8$ mm. Two 2-stage cells of different length ratios L_1/L_2 were simulated, because the length of L_2 should play an important role for the characteristics. Two plane distances, $l = 6$ mm and $l = 15$ mm, are presented. A color map of the two planes is shown for each cell type in Fig. 5.10, where the color codes the film thickness at a given point of the plane.

The Knudsen cell results in a thickness distribution with a central peak and a strong gradient toward the areas of the plane further away from the center. This gradient is less pronounced for the plane at larger distance of the cell. The open cell exhibits a rather sharply delimited central area of relatively homogeneous thickness compared to the Knudsen cell. This area is approximately of the size of the crucible diameter at close distance and widens with increasing distance of the plane. The results of the long 2-stage cell with $L_1 = 15$ mm and $L_2 = 60$ mm are comparable to the open cell. For the shorter 2-stage cell with $L_1 = 45$ mm and $L_2 = 30$ mm one sees a behavior with characteristics from Knudsen and open cell. At 6 mm distance one can see the delimited central area stemming from the *side 2* region, but the inner part is not as homogeneous, rather resembling the distribution of the Knudsen cell. For the larger distance a stronger widening can be seen, compared to the open cell and the long 2-stage, which results purely from the reduced length L_2. It can be summarized that one finds the expected behavior for the 2-stage cell: The introduction of the *side 2* region improves the focusing of the cell compared to a Knudsen cell and analogous to an open cell, the focusing increases with the length L_2. It should be mentioned that the very different total coverage of the planes results from the different simulation times necessary. Simulation time increases with the size of the simulated cell surface and depends also inversely on the size of the relevant orifice.

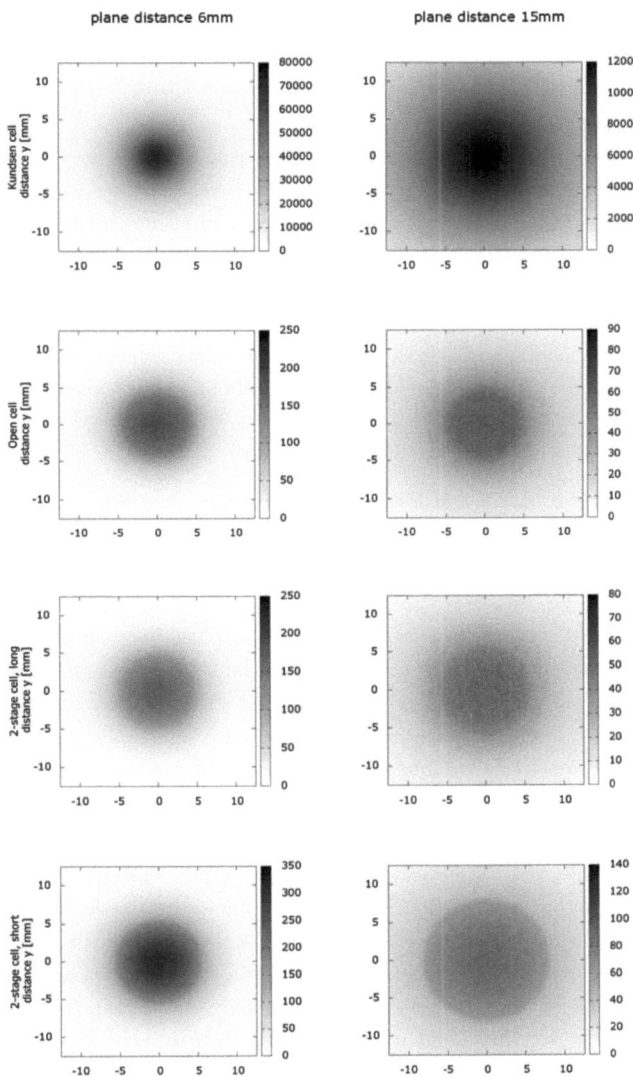

Figure 5.10: Comparison of the different cell types

From top to bottom the simulation for a Knudsen cell, an open cell, a long 2-stage and a short 2-stage cell are shown. On the left side a plane in 6 mm distance from the source is depicted, on the right side the distance is 15 mm. Knudsen and 2-stage cells are simulated in the simplified model. The parameters, according to Fig. 5.6 Knudsen cell $h = 1$ mm, open cell $d = 8$ mm, $L = 75$ mm; long 2-stage $h = 1$ mm, $L_2 = 30$ mm, $d_2 = 8$ mm and short 2-stage $h = 1$ mm, $L_2 = 15$ mm, $d_2 = 8$ mm. The total coverage of the planes varies strongly, which reflects the strong dependence of the simulation time on the simulated surface and the size of the source area. Therefore the long 2-stage cell has the worst statistics.

5.4 Experiments

To verify the results of the simulation a series of experiments with different crucible geometries (cp. Fig. 5.6) was performed.

The samples were prepared in the test chamber (cp. chapter 3). Pentacene was evaporated from an Al_2O_3-crucible equipped with different stainless steel apertures onto glass microscope slides. The thickness of the prepared thin films was measured via optical absorption. The glass slide was mounted on a translation stage with 10 µm resolution and illuminated with a laser with a wavelength of $\lambda = 635$ nm. The laser beam was further collimated with an aperture to reduce the illuminated area to about 0.8 mm^2. In line with laser and glass slide a silicon diode was mounted and the photo current was measured. The microscope slide was moved step by step to measure the absorption position-dependently and obtain a thickness profile of the thin film according to Beer's law

$$I(d) = I_0 \exp(-\alpha \cdot d) \tag{5.11}$$

where d is the thickness of the thin film, α is the absorption coefficient, $I(d)$ the measured photo current and I_0 the current at $d = 0$. This is by definition the current measured, when a clean glass slide is illuminated. The experiment was carried out in a box to minimize the background illumination.

Fig. (5.11) to (5.14) show the simulation and the relative thickness as determined with the absorption measurement for the three cell types. For the 2-stage cell two different cells with length $L_2 = 30$ mm and $L_2 = 15$ mm were investigated. The red curves show the measured thickness distribution, the black curves show the respective simulation. With a red bar it is indicated, where the deposition is limited due to the edge of the sample holder. Beyond the red bar the glass was shaded by the sample holder and no deposition could occur. Arrows are used to indicate distinct features that can be observed in simulation and experiment.

The simulations show good agreement with the experiments, reproducing the indicated features of the respective thickness distributions. The best agreement is achieved for the Knudsen cell, which shows a perfect $\cos^4$-behavior (Fig. 5.11). For the open cell the simulation shows a change of slope at $x = 4$ mm. This can also be observed in the measurement. One can also see that the measurement deviates from the simulation mainly at the center ($x = 0$), where the highest coverage is reached. Here problems with saturation can occur. For the short 2-stage cell the center of the measured distribution is very flat, also suggesting a problem with saturation. For both, short and long 2-stage cell, the changes of slope in the thickness distribution can be matched to the simulation. The overall agreement between simulation and experiment is not as good as for open cell and Knudsen cell.

In general, the results are not satisfactory from an experimental point of view. First of all it proved difficult to clean the glass microscope slides properly, which led to defects or inhomogeneities of the prepared thin films, sometimes visible with the naked eye. Secondly, the roughness of the prepared samples influences the absorption measurement due to diffuse scattering of the laser light. As a third point, reflexion at the front or back of the glass substrate

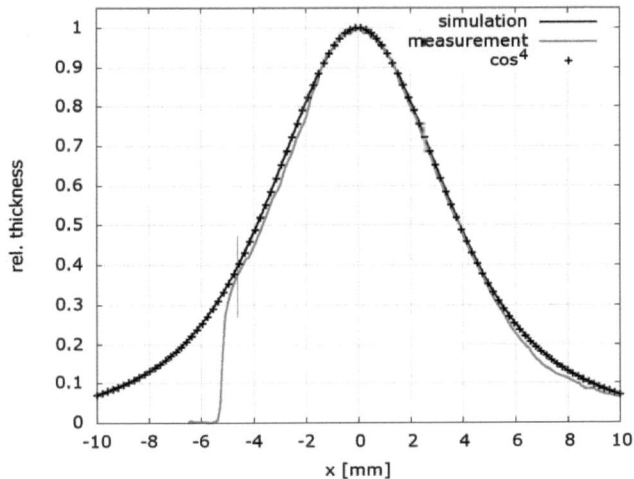

Figure 5.11: Knudsen cell, plane distance 6 mm
The simulation of a Knudsen cell is compared to a measured distribution. Geometric parameters of the cell are $d = 8$ mm, $L = 75$ mm and $h = 1$ mm. For comparison the $\cos^4$-distribution is depicted and is in excellent agreement with simulation and experiment.

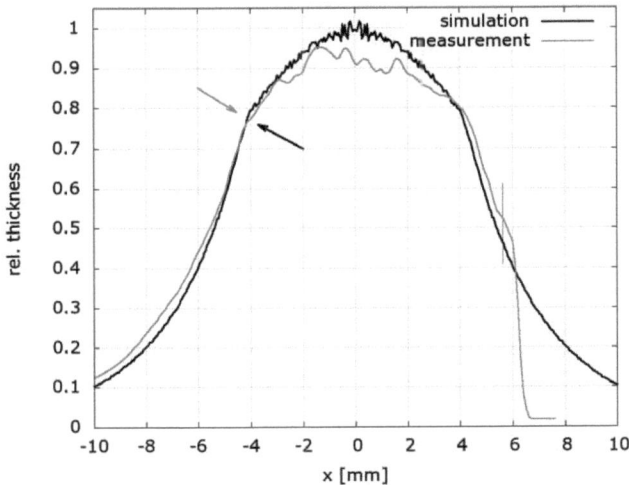

Figure 5.12: Open cell, plane distance 6 mm
The simulation of an open cell is compared to a measured distribution. Geometric parameters of the cell are $d = 8$ mm and $L = 75$ mm. Simulation and measurement show good agreement, except for the central part around $x = 0$. The prepared sample was of minor quality in this region. This is attributed to insufficient cleaning of the microscope slide used as a substrate.

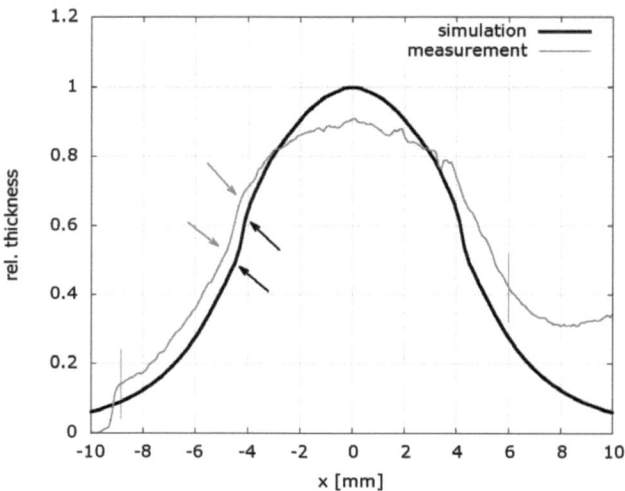

Figure 5.13: Short 2-stage cell, plane distance 6 mm
The simulation of a 2-stage cell is compared to a measured distribution. Geometric parameters of the cell are $d_1 = d_2 = 8$ mm, $L_1 = 60$ mm, $L_2 = 15$ mm and $h = 1$ mm. Characteristic features of the two distributions coincide (see red and black arrows), but the overall agreement of the two curves is not satisfying. This is attributed to a too high thickness of the prepared sample in the central area. Therefore the measured distribution is rather flat in the central part and a relative scaling of the two curves was difficult.

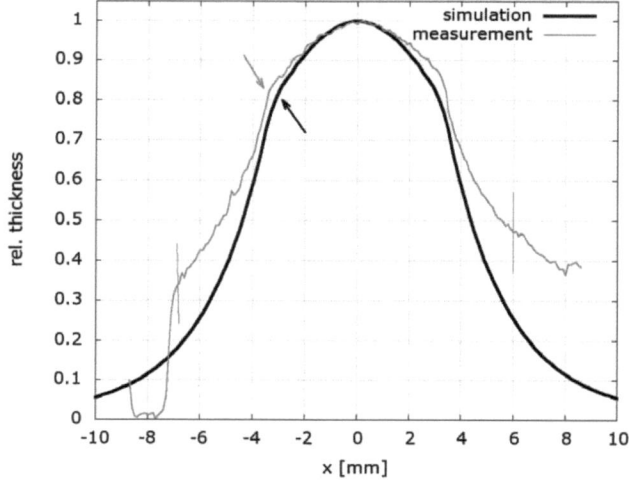

Figure 5.14: Long 2-stage cell, plane distance 6 mm
The simulation of a second 2-stage cell with larger L_2 is compared to a measured distribution. Geometric parameters of the cell are $d_1 = d_2 = 8$ mm, $L_1 = 45$ mm, $L_2 = 30$ mm and $h = 1$ mm. Again, characteristic features of the distributions match, but the overall agreement is unsatisfying. The deviation might partly be caused by an additional deposition onto the prepared area, due to a second deposition experiment on the same microscope slide.

might influence the thickness measurement. For further studies one should therefore search for a different method to measure the thickness. First tests with an alpha-stepper showed promising results, but only a model with 10 mm measuring distance was available, which was not sufficient to thoroughly study the prepared samples.

5.5 Collisions

As mentioned before, collisions are necessary to simulate a real Knudsen cell. The simulations shown until now all used the simplified model, omitting the collisions. To demonstrate the deficiency of the program without collisions, a simulation of a full Knudsen cell is shown in Fig. 5.15. The diameter of the cell and the size of the orifice were fixed to $d = 8$ mm and $h = 1$ mm, respectively, and the length of the cell was varied from $L = 5$ mm to $L = 50$ mm. One can see that the agreement with the $\cos^4$-dependence decreases with increasing length of the cell. This can be understood with the help of the sketch on the right side. The area on the plane in direct sight from the base of the crucible through the crucible becomes smaller and smaller with increasing length of the cell. This area marks roughly the region, where the $\cos^4$-behavior is expected to hold, because of a simple pinhole camera effect. This pure geometric effect has also been verified with a *cold wall* simulation[1], meaning the crucible walls were kept at 0 K to single out the contribution of direct evaporation from the source area. Table 5.1 shows the calculated radius $x(L)$ of the area on the plane as a function of the length of the cell.

L [mm]	5	10	20	30	50
x [mm]	4.5	2.25	1.12	0.75	0.45

Table 5.1: Direct beam cut-off

For a cell of length L the geometric cut-off of the direct beam on a plane of distance 5 mm is calculated.

The simulation demonstrates the expected behavior, namely, that a pure molecular flow simulation cannot provide a correct result for the Knudsen geometry, and serves as a comparison for the simulations with collisions.

Two collision models have been investigated which are named mean free path model and Maxwell Boltzmann model after the basic idea the respective model is based on. In the following sections the models are shortly introduced and a comparison of the results is given.

5.5.1 Mean free path model

The basic idea of this first model is to define a mean free path λ, which can in principle be derived from the pressure inside the Knudsen cell. Collisions occur with a probability of $P(x_c) = e^{-x_c/\lambda}$. For every evaporated particle a collision length x_c is determined by $x_c = -\lambda \cdot \ln(r)$, where r is a random number. When the path length l to the next piercing point with wall or orifice

---
[1]The results of this simulation are not shown here.

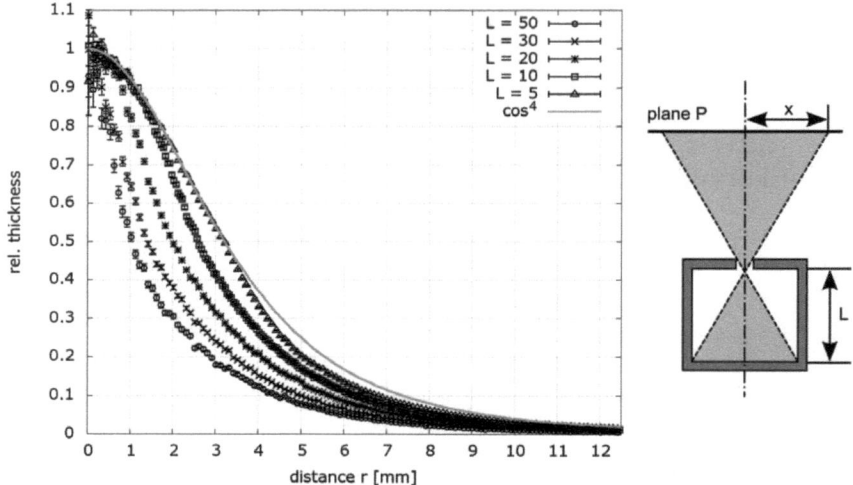

Figure 5.15: Simulation of a Knudsen cell without collisions
A Knudsen cell with $d = 8$ mm, orifice $h = 1$ mm and variable length $L = 5, 10, 20, 30, 50$ mm was simulated. The data shown refers to a plane distance of 5 mm. The ideal $\cos^4$ law is also depicted and the simulated distribution deviates from this behavior with increasing length of the cell. The reason for this is shown in the right sketch. With increasing length of the cell a smaller area of the plane is reached from particles evaporating from the source at the base of the crucible. This portion contributes to a significant part to the distribution, which can also be verified by simulating a *cold wall* crucible.

is shorter than x_c no collision occurs. If $l > x_c$ a collision takes place at $\vec{s} = \vec{p} + x_c \cdot \vec{d}$, where $\vec{p}$ is the particles position vector and $\vec{d}$ the particles direction vector. The particle position is then set to $\vec{s}$ and a new direction vector $\vec{d}$ is drawn from an isotropic distribution. Now a new collision length is determined and the process is iterated until the particle hits the crucible wall or leaves the cell. To avoid infinite run time a certain cut-off for the number of allowed collisions must be inserted.

To monitor the collision process additional output files have been created.

lambda-ideal file Here all assigned collision lengths x_c are stored in a histogram and the mean free path can be calculated as a comparison to the inserted value.

lambda-real file Since the collision is only executed if the total path length of the evaporation step is $l > x_c$ the actual *used* x_c values may differ significantly from the ideal distribution, depending on λ and the cell geometry. In this file only the x_c's for actually executed collisions are stored and an effective mean free path can be derived.

scatter number file To get an idea about the influence of the cut-off value for the number of collisions the number of executed collisions for each particle is stored in a histogram.

The mean free path model is analogous to the description of Kirsch and Gericke [60, 61]. They studied cylindrical and conical open cells experimentally and with a Monte Carlo simulation based on the trajectory model. They observe a broadening of the thickness distribution with decreasing λ/increasing pressure and obtain very good qualitative agreement of experiment and simulation. With our simulation we can reproduce these results for an open cell (cp. Fig. 5.16).

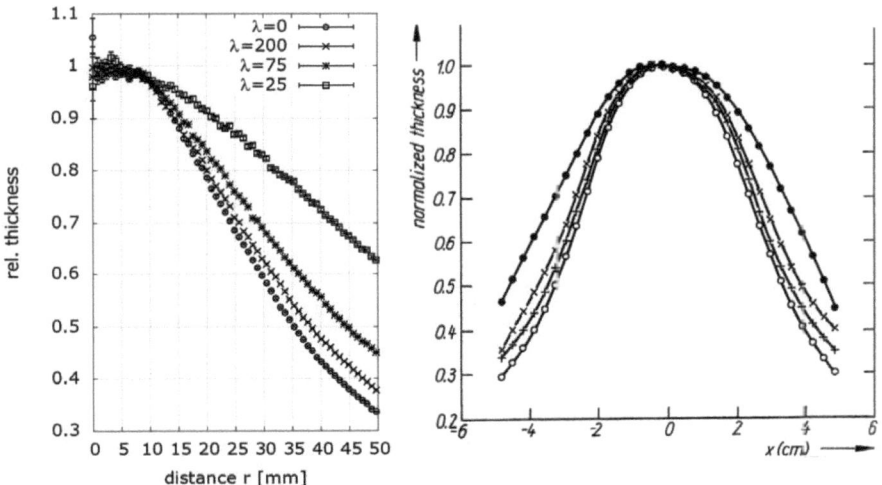

Figure 5.16: Open cell with collisions

On the left hand side the simulation of an open cell with collisions based on the respective mean free paths λ is shown. $\lambda = 0$ corresponds to a simulation without collisions. As expected, the distribution on the plane broadens if collisions are taken into account. This effect is more pronounced if λ is small, meaning that more collisions occur.
On the right hand side the simulation results of Kirsch et al. [61] are shown. They have found excellent agreement between their simulation - featuring the same collision model used here - and experiment.

Nevertheless, the ultimate test for the collision model is the simulation of a Knudsen cell and the question whether the cosine law is reproduced. A Knudsen cell with $d = 8$ mm, $L = 30$ mm and an orifice of $h = 1$ mm was simulated using different mean free paths λ. The results are shown in Figure 5.17. $\lambda = 0$ mm corresponds to a simulation without collisions. Two cases have to be distinguished 1) $\lambda > h$ and 2) $\lambda < h$. For 1) the distribution with collisions should approach the cosine law if a correct collision model is used. For 2) one aspect of the definition of a Knudsen cell is no longer fulfilled, namely, that no collisions occur inside the orifice and therefore a different behavior is expected. This is exactly what is seen in Fig. 5.17. The simulations with collisions at $\lambda = 1...3$ mm approximate the $\cos^4$-distribution of an ideal Knudsen cell in contrast to the simulation with $\lambda = 0.5$ mm. One should note that - counter intuitively - $\lambda = 3$ mm shows the best agreement with the cosine law. One expects smaller mean free paths to show better agreement, as long as no collisions occur inside the orifice. A second observation is that the simulation with $\lambda = 0.5$ mm provides a narrower distribution than the simulation without collisions. As a first assumption one would always expect a broadening of the thickness distribution with collision in contrast to a collision-free simulation.

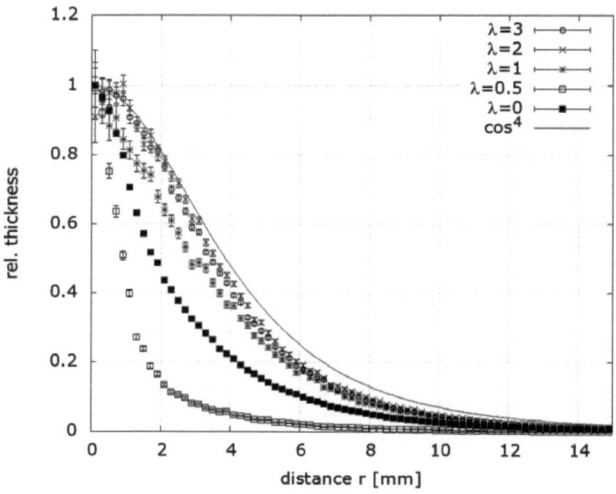

Figure 5.17: Knudsen cell with collisions
The simulations of a Knudsen cell with collisions are shown for different mean free paths λ. Dimension of the cell are $d = 8$ mm, $L = 30$ mm and $h = 1$ mm. For reference a cell without collisions ($\lambda = 0$ mm) and the cosine law are shown. The simulations with $\lambda = 1...3$ mm show good agreement with the cosine law in the central area and deviations at larger r. For $\lambda = 0.5$ mm collisions inside the orifice may occur conflicting with the definition of a Knudsen cell. Therefore deviations from the cosine law are expected.

In conclusion, collisions were implemented correctly using the mean free path model, as the comparison to the results of Kirsch and Gericke shows. Regarding the simulation of a Knudsen cell, improvements compared to simulations without collisions were made, but the overall behavior and especially the dependence on the mean free path length remains unclear. To investigate these open issues a second collision model was implemented.

5.5.2 Maxwell Boltzmann model

The starting point for this second model is the fact that, as shown in section 5.3.11, particles at a given distance from the orifice can only escape if their velocity v exceeds a certain minimum value. This can also be interpreted as a scattering event, where the particle changes direction after $l = v \cdot t$ in analogy to 5.3.11. Practically, the exponential probability distribution in section 5.5.1 is exchanged with a Maxwell-Boltzmann-like velocity distribution $P(v) = c \cdot \sigma \cdot v^2 \cdot e^{\frac{-v^2 \cdot \sigma}{2}}$ with σ and c as free parameters. A velocity is assigned to each particle according to $P(v)$, which corresponds to a collision length $x_{mb} = v \cdot t$, where the time t is chosen to $t = 1$ s for simplicity. This length is then compared to the calculated intersection.

λ_{scale}	λ_{ideal} [mm]	λ_{real} [mm]
3	6	4.7
2	4	3.5
1	2	1.8
0.5	1	1

Table 5.2: Comparision of the different λ's

The table shows the relation of the different λ's used in the simulation. λ_{scale} is the scaling factor inserted in the simulation. λ_{ideal}, the mean free path calculated from the lambda-ideal file, is twice as large as λ_{scale}. λ_{real}, finally, is the *true* mean free path that is calculated from the collision that have actually been executed. λ_{real} is in general equal or smaller than λ_{ideal}, depending on the crucible geometry.

To compare the results of this collision model and the mean free path model the same Knudsen cell was simulated. To see the influence of the mean free path of the particles and related to this, the number of collisions they suffer until escape, the following method was employed: The time t was chosen to $t = 1$ s as in the previous section and a dimensionless mean free path λ_{scale} was used as a scaling factor for the velocity in the Maxwell-Boltzmann distribution. The parameters σ and c were initially adjusted, so that a similar range of mean free paths was covered. The scaling with λ_{scale} results in a shorter mean free path and larger number of collisions for small λ_{scale} and vice versa for large λ_{scale}, in analogy to the definition of a mean free path. The results are shown in Figure 5.18. Since λ_{scale}) is only a dimensionless scaling factor and not directly the mean free path, the mean free path was determined after the simulation from the lambda-real and lambda-ideal files (cp. Table 5.2).

The simulation shows that the thickness distribution is close to the ideal cosine law for all mean free paths larger than the orifice of the cell. The result does not seem to be as sensitive to the exact value of λ as for the *mean free path model*, were larger differences between the simulations of different λ's can be seen. On the other hand there are more free parameters available here (λ, σ, c), which have been arbitrarily chosen to provide reasonable mean free paths. The influence of the individual parameters has not been studied.

5.5.3 Conclusion

Two collision models were tested for the simulation of a Knudsen cell. Both collision models show an improvement of the thickness distribution towards a $\cos^4$ distribution compared to a collision-free model. Also, for both models a change of behavior can be seen if the mean free path is in the order of the orifice of the Knudsen cell, as is theoretically expected. For both models only a small set of parameters has been tested, which does not allow for a definite judgment of the capabilities of the models at this point. At first glance the Maxwell-Boltzmann model shows a faster convergence to the $\cos^4$-distribution, but it has a rather constructed physical basis. In contrast, the mean free path model has a straight forward physical interpretation and an experimental verification [60, 61], but the results for the Knudsen cell are not easily understood. From a theoretical point of view more simulations have to be performed to explore

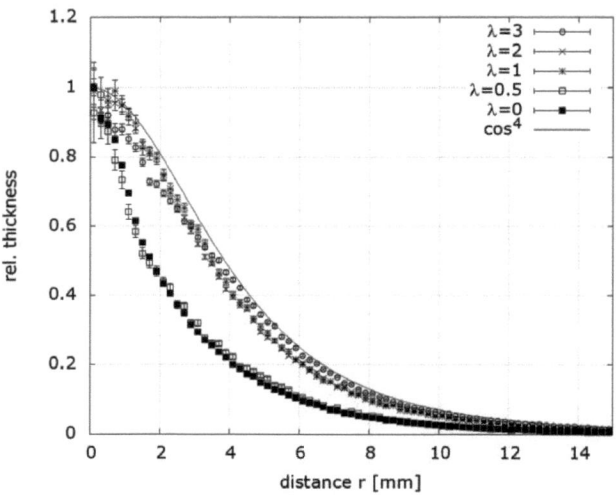

Figure 5.18: Knudsen cell with collisions - 2
The simulations of a Knudsen cell, identical to the one simulated in 5.17 are shown. Dimension of the cell are $d = 8$ mm, $L = 30$ mm and $h = 1$ mm. For reference a cell without collisions ($\lambda = 0$ mm) and the cosine law are shown. This simulation is now based on the *Maxwell-Boltzmann model*. In contrast to 5.17, there is a better convergence of the simulation to the cosine law. Similar to the simulation with the *mean free path model*, when λ is of the size of the orifice, the model breaks down.

a larger parameter space with both collision models to come to a final conclusion. From an experimental point of view it is recommended to use the simplified cell model rather than simulating collisions - whenever possible - for the simple reason of saving computation time. To be able to simulate the intermediate region between molecular flow and viscous flow the implementation of a functioning collision model is of course mandatory and will therefore be beneficial in any case.

Chapter 6

Preparation

In this chapter the preparation of single crystals via solution growth and thin films via evaporation is presented. Although the experimental focus was clearly on thin film preparation, the solution growth experiments were important to obtain source materials for evaporation.

6.1 Single crystal growth

One batch of ET was synthesized by the group of Prof. Dr. Winkler[1], a second batch of ET and a batch of TCNQ were purchased from Alfa Aesar. $Cu(NCS)_2$, $ET_2Cu(NCS)_2$ and (ET)-(TCNQ) are not commercially available and therefore single crystals were grown with different methods of solution growth.

6.1.1 $Cu(NCS)_2$

$Cu(NCS)_2$ [62] was prepared following the protocol of Müller and Ueba [4]. NaSCN was used, instead of KSCN, for the reaction with $CuSO_4 \cdot 5H_2O$.

$$2NaSCN + CuSO_4 \cdot 5H_2O \rightarrow Cu(NCS)_2 + Na_2SO_4 + 5H_2O$$

A dark brown powder was produced and washed with cold water and diethyl ether. Afterwards the product was first dried over H_2SO_4 and then stored in a pumped desiccator over P_2O_5. The powder was studied with X-ray diffraction, with which the successful preparation of $Cu(NCS)_2$ could be proved (cp. Fig. 6.1).

6.1.2 $ET_2Cu(NCS)_2$

For the preparation of the CT salt $ET_2Cu(NCS)_2$ the two methods described in [4], the reflux method and the ultrasonic bath method, were applied.

---

[1] Fachbereich Geowissenschaften, Abt. Kristallographie, Universität Frankfurt am Main

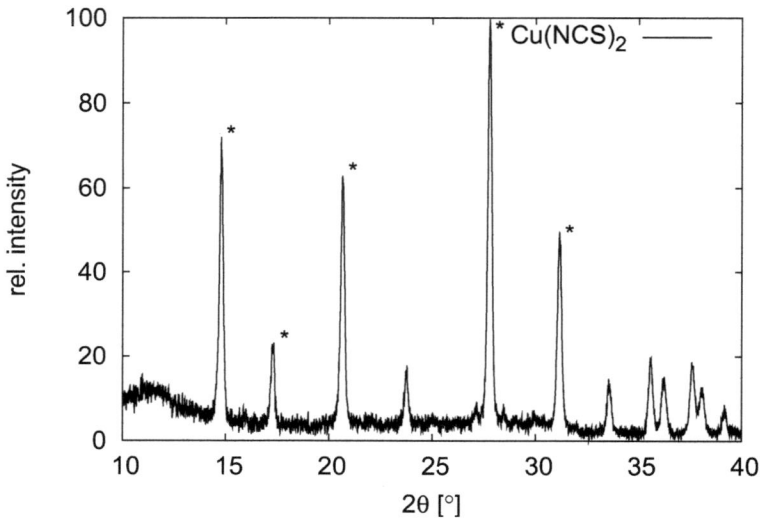

Figure 6.1: Powder spectrum of Cu(NCS)$_2$
Shown is the powder spectrum of Cu(NCS)$_2$ as prepared in section 6.1.1 measured with X-ray diffraction. (*) indicates peaks of Cu(NCS)$_2$ as reported by Hunter et al. [27]. Detailed powder data of the compound was not available.

Ultrasonic bath method

ET and Cu(NCS)$_2$ were added in a stoichiometric 2:1 ratio of 52 mg ET (0.135 mmol) and 12 mg Cu(NCS)$_2$ (0.067 mmol) to 30 ml tetrahydrofurane (THF). The beaker with the mixture was put into an ultrasonic bath. Immediately, a dark brown powder formed at the bottom of the beaker. After 35 min of sonication the powder was filtrated through a suction filter. The very fine brown powder clogged the filter paper prohibiting a filtration. Therefore the remaining crystals in solution were left in the beaker to crystallize. The powder extracted through filtration - I - and the powder after crystallization in the beaker - II - both showed no conclusive X-ray data. Although the rapid color change after switching on the ultrasonic bath suggests the formation of the charge transfer salt, this could not be supported by the X-ray analysis.

Reflux method

109 mg (0.283 mmol) of ET and 26 mg (0.145 mmol) Cu(NCS)$_2$ were suspended in about 50-80 ml of THF in a round flask. The solution was heated to reflux temperature and kept there for about one hour, with occasional shaking of the round flask to dissolve ET which crystallized at the edge of the fluid level. After filtration 55 mg (0.058 mmol) of dark, brown/black powder could be recovered, which corresponds to a yield of 40 %. The X-ray diffraction pattern of the powder is shown in Figure 6.2. The measured spectrum has been shifted 0.3° to coincide with

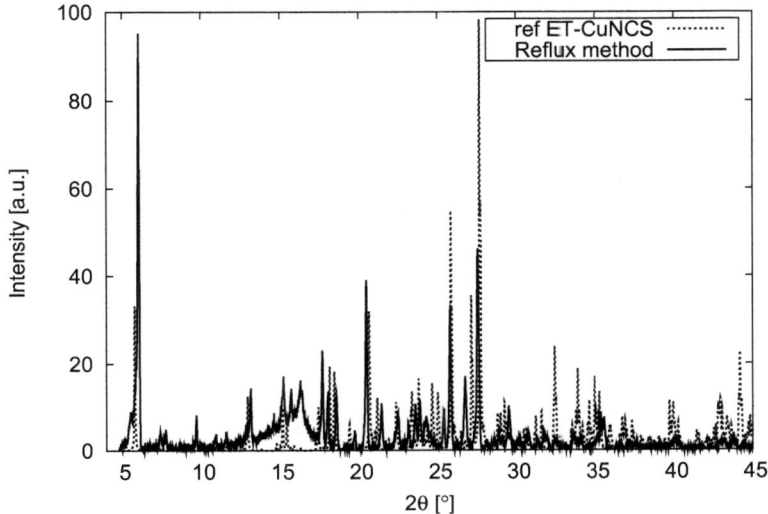

Figure 6.2: Powder spectrum of $ET_2Cu(NCS)_2$
The reference data is taken from [30]. The measured data was shifted 0.3° in 2θ. The powder data is similar to the reference data, when shifted, but the origin of this shift is unclear. It cannot be explained by an error in the height adjustment of the sample in the diffractometer. Furthermore the measurement, including the shift, was reproduced after a completely new alignment of the sample in the goniometer.

the reference data. After the shift, a good agreement is given. A misalignment of the sample can be excluded, because the sample was measured again after a new alignment with the same result. This can not be easily explained, therefore an independent verification of the successful synthesis of $ET_2Cu(NCS)_2$ would be necessary. Since evaporation experiments with the powder were not successful, this was not further pursued.

6.1.3 (ET)(TCNQ)

Solution growth of (ET)(TCNQ) was performed by Prof. Dr. Michael Huth. Three different solvents were used to grow crystals, namely Dichlormethane (DCM), Dichlorethane (DCE) and THF. ET and TCNQ were dissolved into a beaker containing the respective solvent and were moderately heated under stirring on a heating plate. As soon as the crystallites were totally suspended, the two beakers were poured together and left in the extractor hood or a slightly pumped desiccator for slow evaporation. As a result of the lower solubility of ET, the reaction was carried out with an excess of TCNQ. The evaporation of the solvent took from several hours for THF to 2-3 weeks for the other solvents. After evaporation black crystallites of (ET)-(TCNQ) could be recovered. They were mixed with yellow crystallites of TCNQ, as a result of the non-stoichiometric weighted sample. It was not possible to separate the (ET)(TCNQ) fully from the pure TCNQ, therefore TCNQ is a known impurity in the (ET)(TCNQ) source

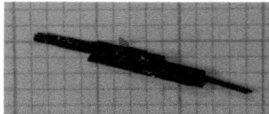

Figure 6.3: (ET)(TCNQ) crystal, grown in DCE
This crystal was the largest crystal obtained by solution growth. The length of the crystal is about 14 mm.

material.

Size and geometry of the (ET)(TCNQ) crystals varied strongly upon solvent and evaporation speed from small platelets to long needles. One of the largest crystals is shown in Fig. 6.3.

Different crystal phases were identified, depending on the solvent used for the solution growth. The respective powder spectra are shown in Fig. 6.4.

- THF: a mixture of monoclinic phase and β'-phase
- DCE: β'-phase
- DCM: monoclinic phase

The β''-phase was not observed.

The ultrasonic bath method was also tested for (ET)(TCNQ), in the hope to speed up the reaction. It was observed that the ET dissolves more easily under the agitation of the ultrasonic bath. The experiment was performed with 50 mg (0.13 mmol) ET and 29 mg (0.14 mmol) TCNQ suspended in a beaker with 30 ml THF. The solution was kept for 30 min in the ultrasonic bath. No reaction could be observed. Afterwards the beaker was kept under the extractor hood to evaporate the solvent, yielding black CT crystals as in the solution growth experiments without use of the ultrasonic bath.

6.2 Thin films

For the preparation of devices such as tunnel junctions or field effect transistors the availability of thin films is a prerequisite. With the method of organic molecular beam deposition it was attempted to prepare crystalline thin films of the various organic materials presented in chapter 2. Since the literature on evaporation/sublimation of these materials is in many cases not extensive, the first step was usually to determine the sublimation temperature of the material by a controlled ramp-up of the effusion cell temperature until a deposition could be detected. In this process the substrate temperature was usually kept at room temperature (25 °C to 30 °C) to minimize re-evaporation from the substrate.

When sublimation and deposition were successful, the growth of the material was studied in detail in dependence on sublimation temperature, substrate temperature and substrate material.

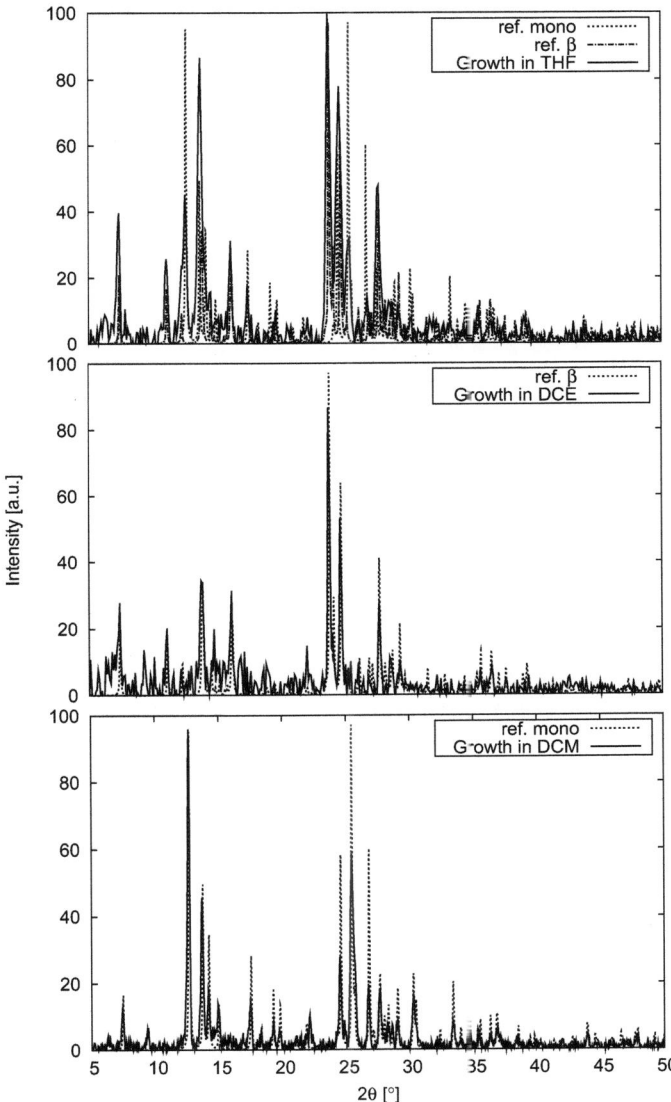

Figure 6.4: Powder spectra of (ET)(TCNQ)
From top to bottom the powder spectra of (ET)(TCNQ) grown in THF, DCE and DCM are shown. For the growth in THF a mixture of monoclinic phase and β'-phase was found, for DCE only β'-phase was observed and for DCM only the monoclinic phase.

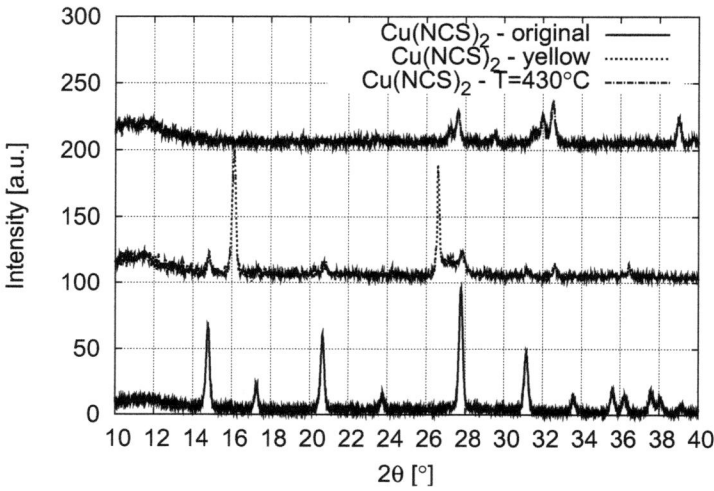

Figure 6.5: X-ray diffraction of Cu(NCS)$_2$
The first curve shows the powder pattern of the original Cu(NCS)$_2$ before heating. The second curve shows the pattern obtained from the yellow precipitate and the third curves shows the pattern of the material after being heated to 430 °C. The curves have been shifted vertically for clarity.

For practical reasons, glass microscope slides were used as standard substrates in the test chamber and Al$_2$O$_3$ substrates were used in the other chambers. Al$_2$O$_3$ was chosen, because it is a chemically inert substrate, which has been studied intensively for the growth of metallic as well as organic thin films. It can be used at substrate temperatures much higher than the typical sublimation temperatures of organics, which usually do not exceed 500 °C. Additionally, it is suited for low temperature studies in a cryostat.

For several materials a first sublimation experiment was carried out with a free filament cell and in a second experiment the mini cell was used. Thereby, it was ruled out that the temperature gradient in the free filament cell was the cause of the unsuccessful evaporation experiments.

6.2.1 Cu(NCS)$_2$

Sublimation experiments carried out by Hunter [27] and Ptaszynski [28] in air show a decomposition of Cu(NCS)$_2$ in the presence of water. In reverse, Cu(NCS)$_2$ could be sublimable under ultra-high vacuum conditions, if the source material is sufficiently water-free.

The sublimation of the prepared source material was studied in the test chamber using two types of cells, a free filament cell and a mini cell. The first experiment was carried out with a free filament cell. Although the source material had been dried directly after preparation (cp. section 6.1.1), it was attempted to dry the material in the vacuum chamber before the actual sublimation. For this purpose the cell was kept at 60 °C over a period of 72 h. The temperature was then increased in 10 °C-steps. Starting at 80 °C the brown Cu(NCS)$_2$ powder turned yellow. The temperature was further increased up to 120 °C to exceed the evaporation temperature of water. The color change indicates a reduction from Cu(II) to Cu(I), where the yellow color points to the β-phase of Cu(I)NCS. The chamber was opened and the crucible was taken out to study the source material. The yellow precipitate was superficial and could be removed mechanically. The removed material was studied with X-ray diffraction. The result in comparison to the original Cu(NCS)$_2$ can be seen in Fig. 6.5. Several peaks of Cu(NCS)$_2$ have vanished or decreased in relative height and new peaks have appeared. This confirms that the yellow precipitate is a different crystal phase. Whether or not it is β-phase of Cu(I)NCS cannot be concluded unambiguously.

Assuming that this partial decomposition still stemmed from residual water in the source material, the sublimation experiment was repeated. Again, starting at 80 °C, the material turned yellow. To exclude that the source material had been newly contaminated with water during the mechanical removal after the first experiment, the crucible was kept under vacuum in the second experiment, leaving the yellow precipitate as an impurity in the source material. The temperature was successively increased to 430 °C with only minimal deposition, which can be attributed to impurities. Starting from 290 °C, the source material darkened in color. The experiment was stopped after reaching 430 °C and the source material was taken out. It had changed color to brown again and had baked together to a hard cluster. Under X-ray diffraction the material showed further degradation (cp. Fig. 6.5). The sublimation experiments were therefore concluded as unsuccessful. At a later stage a new attempt was carried out using the mini cell. The source material was for this experiment mixed with quartz sand to prevent agglutination of the source material. The same degradation behavior could be observed.

Consequently, since Cu(NCS)$_2$ could not be evaporated, the method of co-evaporation of ET and Cu(NCS)$_2$ could not be applied.

6.2.2 ET

Sublimation of ET has already been reported by Nollau et al. [24], Miura et al. [22] and Molas et al. [23]. One remarkable point concerning the available publications is that the reported sublimation temperatures vary from 200 °C as reported by Molas to 80 °C as observed by Miura, who claims that decomposition occurs already at 130 °C. According to this contradicting information the sublimation point of ET was practically unknown at the start of the evaporation experiments. Various evaporation series have been conducted under different conditions. Only a short summary will be given here.

Small MBE-system The first experiments were carried out in the small MBE-system, where the sample-cell distance was about 12 cm. In the first evaporation test with the ET provided by Prof. Dr. Winkler the temperature was gradually increased from 70 °C to 170 °C. The cell was finally kept at 150 °C over night, where a sudden pressure increase to $p > 10^{-5}$ mbar occurred. The recovered source material had turned black, suggesting decomposition of the material.

In a second series of experiments with fresh source material, samples were prepared with cell temperatures from 80 °C to 130 °C on Al_2O_3 a-plane substrates. Deposition times were typically between 1 h and 2 h. The substrates were mostly optically clean after the deposition process, with sporadic tiny black crystallites on particular samples. Neither X-ray diffraction nor X-ray reflectometry showed any signs of deposition.

The source material was then exchanged to ET purchased from Alfa Aesar. A series of experiments was conducted with temperatures from 80 °C to 165 °C on Al_2O_3 a-plane substrates with deposition times of 1 h to 3 h. The results were similar to the previous series, again sporadic black crystallites were found and X-ray measurements provided no results. After this series the source material had also turned black [2].

The next step was a modification of the effusion cell. The cell was now equipped with a stainless steel lid with an orifice of 1 mm diameter, which will in the following be called *Knudsen lid*. The deposition temperature was varied from 120 °C to 150 °C. Again no deposition was detected on the substrate. After 7 days at 120 °C with closed cell shutter, the shutter displayed a distinct orange spot. The source material was still orange and therefore assumed to be intact.

OMBD The next experiments were conducted in the OMBD chamber using a cell with a 2 mm Knudsen lid. Sample-cell distance was 25 cm. The quartz microbalance was used to monitor the deposition process, but no rate could be detected. The cell temperature was varied from 60 °C to 130 °C.

Again deposition could only be demonstrated on the cell shutter, not on the substrates. The source material had again darkened in color, even though the temperature had only been increased up to 130 °C.

Test chamber A series of experiments was conducted in the test chamber. The sample-cell distance here was much smaller compared to the previous experiments, varying from 3 mm to about 15 mm. It was intended to find an optimum crucible geometry for evaporation (cp. Chapter 5). The advantage of the test chamber is that the deposition process can be directly observed through a window, when microscope slides are used as substrates.

It was found that ET can in principle be evaporated with all geometries (open cell, Knudsen cell and 2-stage cell), but a closed cell (Knudsen or 2-stage) seems to result in higher evaporation rates. This conclusion is based on the optical thickness of the prepared samples and could not be quantified at that time. A second very important result was, that ET desorbs in vacuum. A prepared sample was left in the chamber over night and the deposition had mostly vanished on the next morning.

Summary From these experiments the following conclusions can be drawn:

1. The evaporation rate of ET is extremely low.

---

[2]The X-ray measurements were performed on a Siemens D 500. Unfortunately, the diffractometer has two background peaks at $2\theta = 8.5°$ and 16.9 degree, which might obliterate the ET $0ll$-reflections. The two peaks have been observed in pure substrate and pentacene measurements and were therefore identified as background.

2. ET desorbs from the substrates in vacuum.

3. Sublimation occurs close to the point of decomposition.

Point (3) prohibits further increase of cell temperature to increase evaporation rates, so the rate can only be compensated by short sample-cell distances and long deposition times. Points (1) and (2) together explain, why no deposition was detected on the substrates prepared in the small MBE and the OMBD chamber - the desorption canceled the small amount of evaporated material arriving at the substrate for these long sample-cell distances. The deposition on the shutter suggests that ET is stabilized once a critical density or sizable seed is accomplished, which was reached on the shutter, but not on the substrate.

The darkened ET was studied with XRD several times to verify, whether decomposition had occurred. When the "black" ET was prepared for powder diffraction with a mortar it was found that the darkening was superficial and that the crystallites were still orange inside. The X-ray diffraction pattern was identical to the pattern of "fresh" ET.

MALDI measurements To verify or disprove the suspicion of decomposition of the ET molecules under sublimation, MALDI measurements have been carried out. As a reference, ET as purchased was investigated. A peak at 384 − 385 u was observed, which is consistent with the molecular weight of 384.7 g/mol. The spectrum is shown in Fig. 6.6.

For the thin film samples a second signal was found at 296 u besides the peak at 384 − 385 u. An example of a spectrum of evaporated ET is shown in Fig. 6.7. This corresponds to a mass difference of 88 u, which can be explained by a separation of two carbon and two sulfur atoms from the ET molecule. A possible breaking point is indicated by the arrows in Fig. 6.8. The open bonds would have to saturate with hydrogen to provide the result of 296 u.

The attempt to verify these results in a second series of experiments was unsuccessful in the sense that both peaks could repeatedly be detected in samples containing *evaporated* ET, but the peak at 296 u could also be found in measurements of samples *not* containing ET. Most probably this was caused by a contamination of the matrix or the used solvents, but it can hence not be excluded, that the signal in the ET measurements is also caused by impurities or a background signal. In this second series, a sample with ET as purchased was not included.

OMBD with different crucibles To exclude that a reaction with the crucible material causes the problems of ET evaporation, three different crucibles were tested: Al_2O_3, graphite and quartz glass.

A series of 15 experiments using a graphite crucible was performed. An overview over the prepared samples is shown in Appendix C, Table C.1. The samples exhibited no peaks under X-ray diffraction. On optical micrographs (cp. Fig. 6.9) small crystallites were visible. The crystallites exhibited very different shapes and colors under the optical microscope, both, on only one sample, but also comparing different samples.

After the completion of this series a quartz glass crucible was used and fresh source material was mixed with sand. Two samples were prepared with deposition times around 5 h. Again,

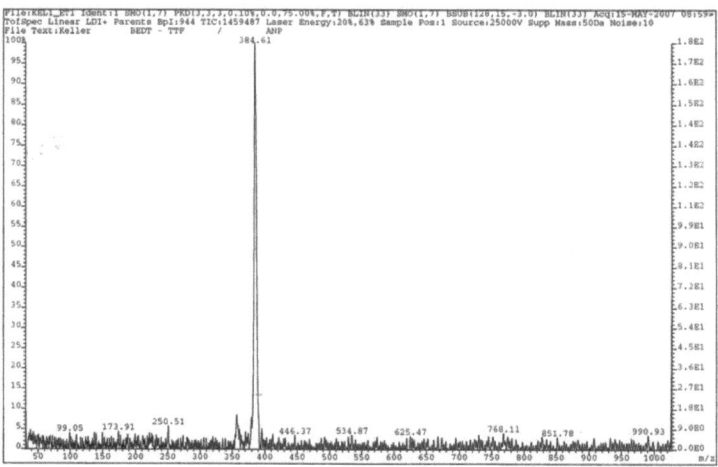

Figure 6.6: MALDI spectrum of ET as purchased

For the sample with ET, as purchased, only a peak at 384 − 385 u is observed. The matrix used in this experiment was DHB (2,5-Dihydroxibenzoic acid).

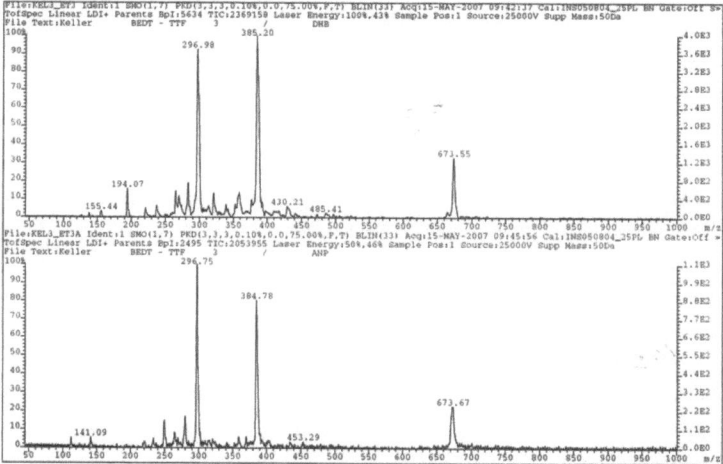

Figure 6.7: MALDI spectrum of evaporated ET

Two experiments with evaporated ET are shown. ET was evaporated onto a glass substrate and then scratched from the surface with a scalpel. Two different matrices were used, DHB and ANP (Aminonitropyridine). For both one can see the peak at 384 − 385 u, as well as a peak at peak at 296 u. This strengthens the suspicion of decomposition of ET.

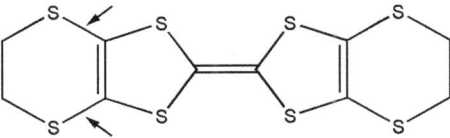

Figure 6.8: Suggested decomposition of ET

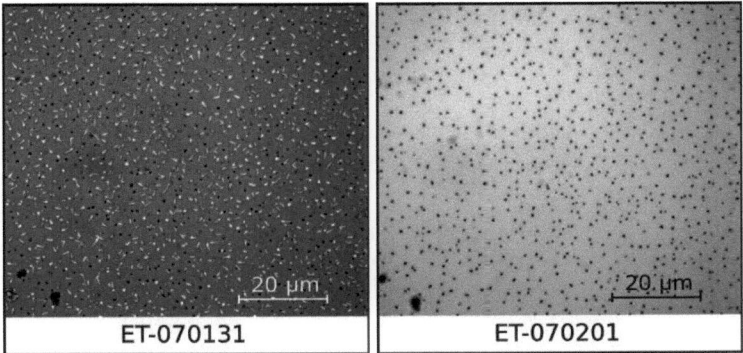

Figure 6.9: Optical micrograph of two ET samples
Two samples are shown to demonstrate the typical morphology of ET samples. Small crystallites of different forms such as cubes or elongated needles are observed. Furthermore, for the sample from the 31.01.07 it seems as if two types of crystallites are present.

X-ray diffraction provided no peaks.

From a later perspective one can assume that evaporation times and also the measurement time for the X-ray diffraction were too short to give conclusive results.

Test chamber with mini cell Samples were prepared in the test chamber using the mini cell. The source material was mixed with quartz sand. A thick orange layer was deposited onto a microscope slide. With the optical microscope orange crystallites are visible as shown in Fig. 6.10. XRD of the sample showed various peaks of ET, similar to the powder pattern (cp. Fig. 6.11). This means that the crystallites do not have a preferred orientation on the glass substrate.

Conclusion

The evaporation of ET proved to be difficult. The evaporation rates were in general low, resulting in typical deposition times of several hours. The samples were without exception microcrystalline and no closed layer of ET could be prepared.

With XRD generally no peaks could be identified, suggesting that the crystallites are oriented randomly on the surface. One exception is the sample prepared in the test chamber using the mini cell. Here, several peaks could be identified showing good agreement to the

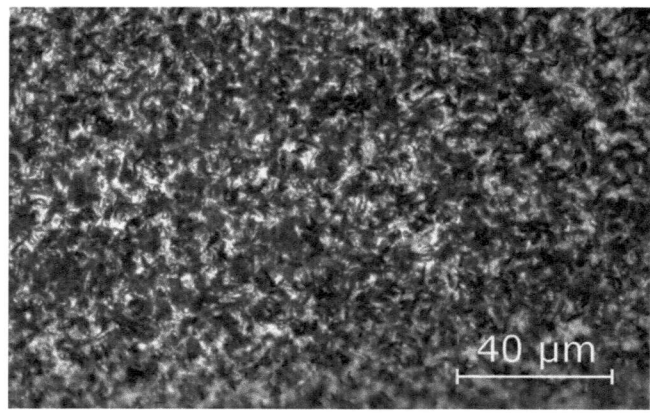

Figure 6.10: Optical micrograph of ET on glass
The deposit is very thick and orange crystallites are observed growing on top of each other. Despite the thickness of the film, it seems as if a closed layer is not reached, yet.

powder data of ET (cp. Fig. 6.11), which proves the random orientation of the crystallites on the substrate.

The color change of the source material to black and the suspected decomposition could not unequivocally be explained nor confirmed.

It has already been pointed out by Knoll et al. [22] that the presence of ET-fragments prohibits the growth of an ordered ET thin film as prepared in [23]. Therefore one can conclude that either impurities in the source material or the suspected decomposition of the material prevent growth of an ordered thin film. This conclusion is strengthened by the successful deposition of ET from (ET)(TCNQ) single crystals presented later in this chapter.

6.2.3 TCNQ

First evaporation experiments with TCNQ were carried out in the test chamber. Between 80 °C and 105 °C evaporation of impurities was observed. At 115 °C evaporation of TCNQ set in. The experiments were continued in the OMBD chamber. There, the evaporation could be monitored with the quartz microbalance. The rate and thickness are hereby only relative values, since the QMB was not calibrated to the organic molecules, but the settings were optimized to maximum sensitivity to obtain a high resolution. An overview over the prepared samples can be found in Appendix C, Table C.2.

It should be noted that the evaporation rate of TCNQ is not stable. It increases if the temperature is increased, but at a fixed temperature the rate decreases from sample to sample. Between the 22.01.07 and the 05.04.07 the OMBD chamber was opened and the TCNQ cell was removed. It was found that the TCNQ had baked together in the crucible forming a hard crust on top of the source material. The source material was thereafter exchanged and the fresh

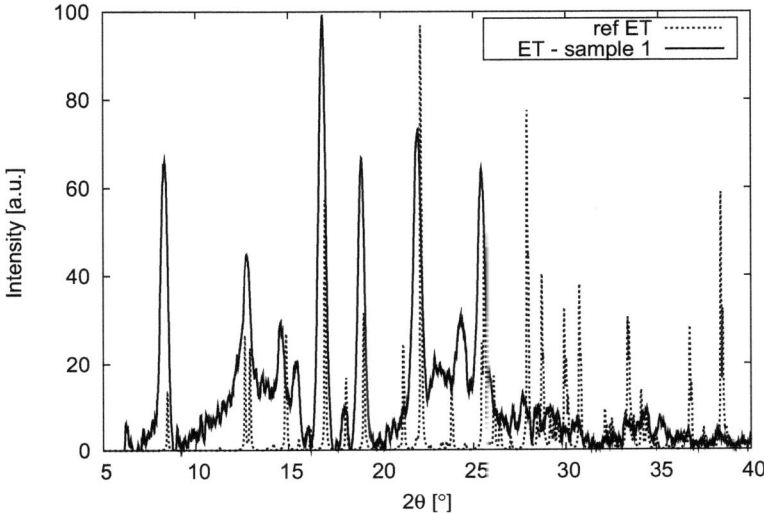

Figure 6.11: X-ray diffraction of ET on glass
The black curve shows the measured X-ray data (background subtracted for clarity). The red curve is the literature data of ET after Kobayashi et al. [25]. The two are in good agreement, proving that the orange crystallites are ET. The crystals are randomly distributed without a preferred orientation.

TCNQ was mixed with quartz sand to avoid this agglutination.

The samples displayed yellow color, as is expected for TCNQ. With X-ray diffraction and reflectometry no results were obtained.

6.2.4 (ET)(TCNQ)

Three different methods were employed to prepare (ET)(TCNQ) thin films:

- co-evaporation of ET and TCNQ,
- evaporation from (ET)(TCNQ) single crystals and
- co-evaporation from (ET)(TCNQ) single crystals and TCNQ.

The reasons for the study of the tree different methods and the results are summarized briefly.

Co-evaporation

As a first step co-evaporation of ET and TCNQ was done in the OMBD chamber. The respective source material was mixed with quartz sand to prevent agglutination of the material in the crucible. ET was evaporated from a quartz glass crucible and TCNQ from an Al_2O_3 crucible.

To maximize the deposition rate for ET at the sample position a free filament cell was modified. The cell was extended to achieve short sample-cell distance of about 2 cm and

No.	Date	T_{et} [°C]	T_t [°C]	T_s [°C]	t [s]	d [kÅ]	sub
A00	15.01.07	120	115	RT	10800	1.31	Al_2O_3 r
A0	01.03.07	150	115	RT	>3800	-	Al_2O_3 r
A1	12.04.07	135	120	30	24550	-	Al_2O_3 r
A2	13.04.07	135	120	80	27000	2.54	Al_2O_3 *
A3	16.04.07	135	120	100	27000	604	Al_2O_3 a/r
A4	17.04.07	135	120	60	27000	1.53	Al_2O_3 a/r
A5	18.04.07	135	120	30	91650	4.36	Al_2O_3 a
A6	23.04.07	135	120	-	27000	4.73	Al_2O_3 r

Table 6.1: Co-evaporation of ET and TCNQ
T_{et} is the temperature of the ET-cell, T_t the temperature of the TCNQ-cell and T_s the substrate temperature. t signifies the deposition time and d is the relative thickness of TCNQ as displayed by the QMB. a and r stand for the respective cut of the substrate. (*) orientation unknown.

mounted on a movable bellows for exact positioning. Because of the inner diameter of the bellows, the outer diameter of the cell had to be reduced. This was achieved by reducing the design to only one radiation shield.

The evaporation process was monitored with the quartz microbalance as well as the pressure gauge. One should note that the quartz microbalance can only receive flux from the TCNQ-cell, because of the close positioning of the ET cell to the sample. An overview over the prepared samples is given in Table 6.1. As already observed for TCNQ, the relative thickness as displayed by the QMB varies strongly, even under nominally equal conditions of cell temperature and deposition time. Reasons for this could be

- impurities, which usually lead to higher deposition rate for the first couple of samples,

- inhomogeneous temperature of the effusion cell, resulting in different rates depending on the filling level or

- agglutination of the source material, which decreases the deposition rate with time.

Agglutination should be prevented by mixing quartz sand into the source material, although it has not been studied which amount of sand is necessary.

There is no clear trend for the change in deposition rate, meaning that several aspects must play a role. Since TCNQ is more volatile than ET and more easily sublimable, it was aimed to offer "sufficient" TCNQ in the following experiments. This was practically ensured by offering an excess of TCNQ. In analogy to the solution growth experiments it was assumed that the growth of the CT salt is possible with an excess of TCNQ.

The samples were investigated with X-ray diffraction, with the following results:

- Sample A1 has a broad peak at $2\theta = 14.8°$. This peak will recur for several other samples.

- Sample A2, A3 and A4 exhibit no features in X-ray diffraction.

- Sample A5 has a peak at $2\theta = 15.5°$ with a FWHM of $0.36°$ and higher order reflection at $31.3°$ (cp. Fig. 6.12). Candidates for these peaks are (110) and (220) of ET with literature

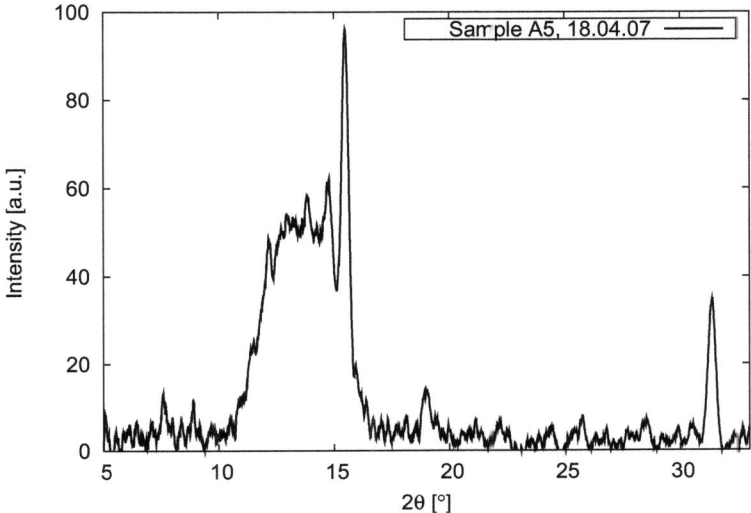

Figure 6.12: X-ray diffraction of sample A5
One can see the peaks at $2\theta = 15.5°$ (ET (110)) and $2\theta = 31.3°$ (ET (220)). Furthermore a broad peak at around $2\theta = 13.7°$ is visible.

values of $2\theta = 15.55°$ and $2\theta = 31.4°$. Further possibilities are $(0\bar{1}1)$ and $(0\bar{2}2)$ of the β'-phase, or $(\bar{3}01)$ and $(\bar{6}02)$ of the monoclinic phase. Position and relative intensity of the peaks suggest ET with a $(ll0)$ orientation, but since these peaks were only observed for one sample no definite conclusion is possible.

Additionally, a broad peak, centered at $2\theta = 13.7°$ is visible.

- Sample A6 shows a peak at $2\theta = 8.46°$ and $2\theta = 16.9°$ (cp. Fig. 6.13). This points to ET in a $(0ll)$ orientation. In a later experiment a well-ordered growth of ET in $(0ll)$ orientation was found (cp. section 6.2.4). Again, a broad peak is observed, here centered at about $2\theta = 13°$.

In conclusion no clear evidence for charge transfer on the substrate under co-evaporation of ET and TCNQ was found. Mostly, no phase could be detected with X-ray diffraction, one sample showed a deposition of ET and for a second sample, Sample A5, the phase could not unambiguously be identified. As a general problem neither the evaporation of ET nor the evaporation of TCNQ is stable and reproducible. This makes it hard to identify adequate growth conditions.

Evaporation from single crystals

To overcome the problems of the ET-sublimation, evaporation from single crystals of the preformed CT-salt is an alternative. For the TTF-TCNQ-charge transfer complex sublimation

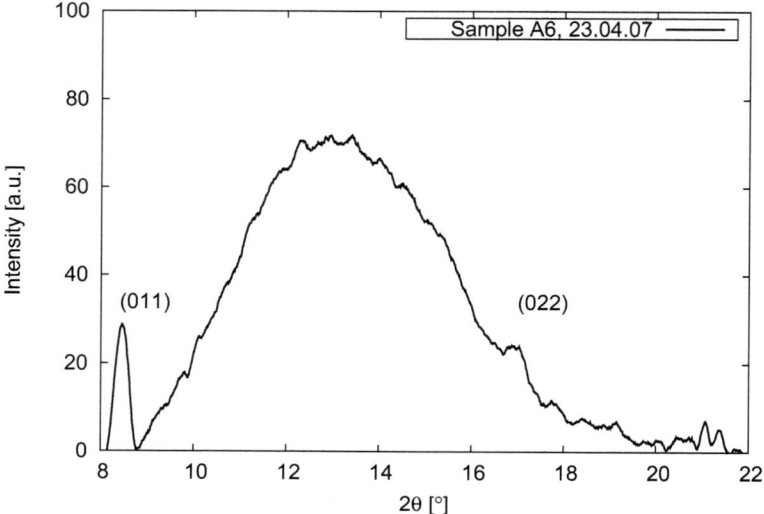

Figure 6.13: X-ray diffraction of sample A6
The first and second order of ET (0ll) and a broad peak at $2\theta = 13°$ are observed.

from single crystals has already been reported [63, 64]. This serves as a proof of principle. Since ET is a larger molecule than TTF one expects a higher sublimation temperature for (ET)(TCNQ) compared to TTF-TCNQ.

Test chamber

First experiments were carried out in the test chamber using the mini cell. Single crystals of (ET)(TCNQ) were powdered, mixed with quartz sand and filled into a glass crucible. Monoclinic phase and β'-phase were mixed together to achieve a reasonable amount of source material. The amount of (ET)(TCNQ) was 27 mg. Three samples were prepared on microscope slides.

Sample 1 The cell was heated from room temperature to 170 °C over the course of five days. Between 160 °C and 170 °C deposition of a yellow film was observed. This is the residual TCNQ. The optical micrograph (cp. Fig. 6.14) shows yellow crystallites, together with black crystallites and XRD shows peaks of TCNQ (Fig. 6.15).

Sample 2 The second sample was deposited at 170 °C-210 °C, over the course of 24h. The optical micrograph (cp. Fig. 6.16) shows black, shiny crystallites. It appears as if two types of crystallites are visible (long needles and small meander-like structures), which seem to differ in geometry and in color. The different color might be an effect of focus or illumination of the microscope, or due to the surface morphology of the crystal. The results of X-ray diffraction can be seen in Fig. 6.17. The measured data was shifted +0.2° to coincide with the powder

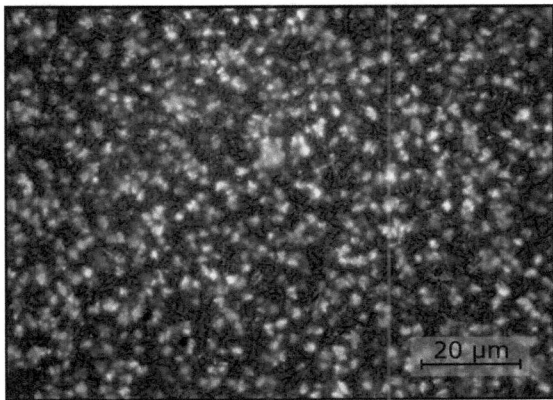

Figure 6.14: Optical micrograph of Sample 1 of (ET)(TCNQ)
In the foreground yellow crystallites are visible. In the background a meander structure of dark crystallites can be seen. Together with the results of XRD (cp. Fig. 6.15) one can conclude that the yellow crystallites correspond to TCNQ.

data of ET. The origin of the shift remains unclear. It cannot be caused by a height error during mounting of the sample, because this results in a shift depending on the angle, not a constant shift. A strained lattice of the crystallites can also be excluded, because this should effect different lattice orientations in a different way. A possible explanation is an error in the adjustment of the X-ray diffractometer, probably a misalignment of the ω circle. Since the data matches the powder data of ET very nicely after the shift, it is believed that the shift is caused by a misalignment rather than being a property of the sample.

Sample 3 A third sample was prepared with deposition at 210 °C for 9 h. The optical micrograph as well as the result of X-ray diffraction are similar to the findings for Sample 2. Sample 3 therefore serves as a confirmation of the results.

At this time it was falsely believed that Sample 2 and Sample 3 show peaks of (ET)(TCNQ). Therefore the experiments in the test chamber were "successfully" concluded. This misjudgment was based on incomplete information on the powder diffraction data for the different phases of (ET)(TCNQ).

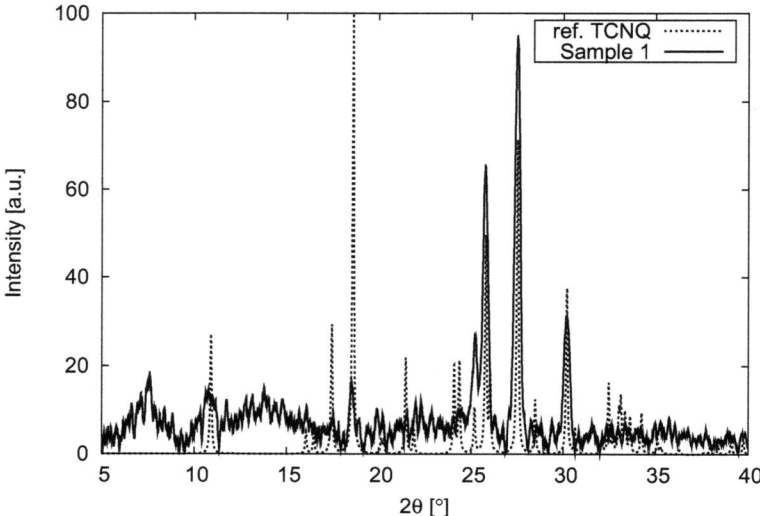

Figure 6.15: X-ray diffraction of Sample 1 of (ET)(TCNQ)
Several peaks of the measurement of Sample 1 coincide with the powder spectrum of TCNQ. This can be interpreted as a large number of TCNQ crystallites of different orientation with respect to the substrate.

OMBD - pure single crystals

After this first test the experiments were continued in the OMBD chamber. The modified free filament cell was used to provide short sample-cell distances. In analogy to the previous experiment, the single crystals (monoclinic phase and β'-phase) were powdered, mixed with quartz sand and filled into a glass crucible. The total amount of (ET)(TCNQ) was 68 mg.

13 samples were prepared, including 4 samples during ramp up of the cell temperature from room temperature to 125 °C. These four samples served as monitoring samples, helping to determine the evaporation temperature in the OMBD chamber. As mentioned in chapter 4 a significant difference in evaporation temperature has been found for the mini cell in comparison to a free filament cell. This explains why the deposition temperature in the OMBD is 85 °C lower in the subsequent experiments.

The remaining 9 samples were prepared at 125 °C with evaporation times of 7-14 hours. For an overview over all prepared samples compare Appendix C, Table C.3. The substrates used were Al_2O_3 "random", where "random" corresponds to r-plane with a miscut of more than 0.3°.

All samples have microcrystalline morphology with various different features, such as cubic crystallites, hammer-structures, needles and circles. An overview of the various geometries is given in Fig. 6.18.

A first breakthrough was sample B5, prepared on 20.06.07. The X-ray diffraction shows two peaks, which can be identified as the first and second order of $(\bar{1}0l)$ of (ET)(TCNQ), monoclinic

Figure 6.16: Optical micrograph of Sample 2 of (ET)(TCNQ)
Two types of crystallites can be distinguished, black crystallites growing in a meander structure and shiny, elongated needles. It is not clear, whether these two types correspond to different orientations, different crystal structures or chemically different crystallites. The morphology of Sample 3 was comparable.

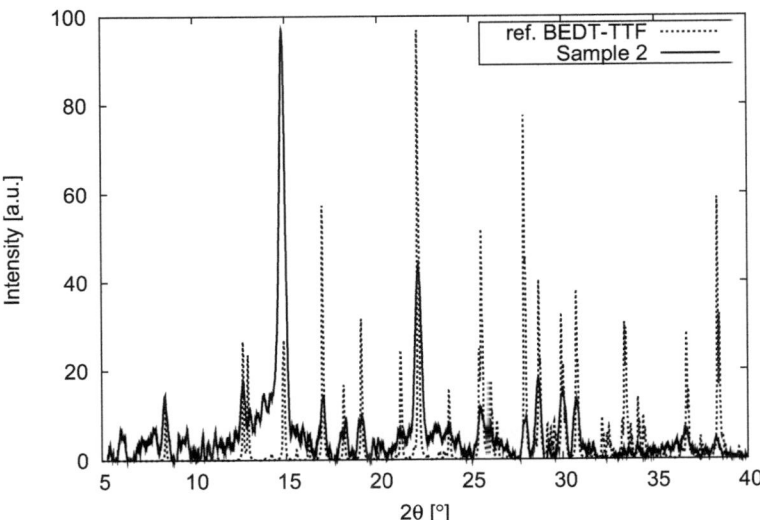

Figure 6.17: X-ray diffraction of Sample 2
The measured diffraction pattern of Sample 2 coincides nicely with the reference data of ET from Kobayashi [25].

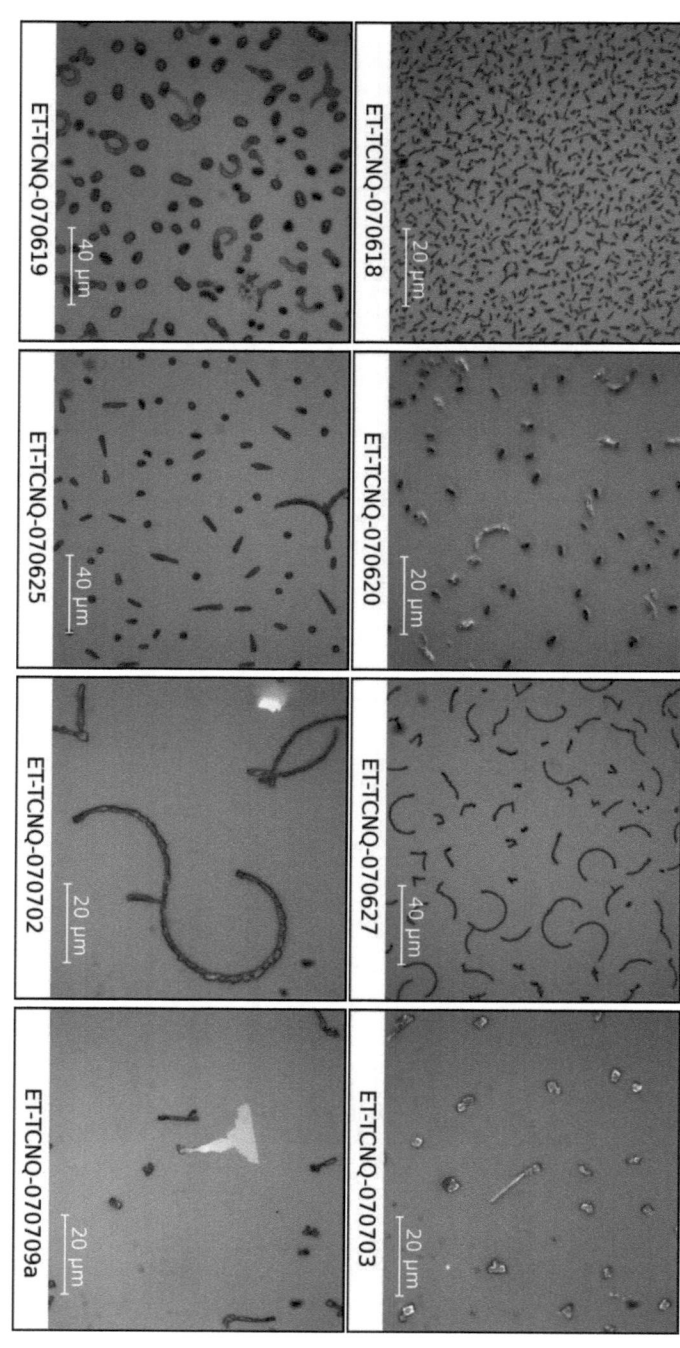

Figure 6.18: Optical micrographs of (ET)(TCNQ) samples

The different pictures are examples for the different morphologies observed for the (ET)(TCNQ) samples. A clear correlation between optical appearance and X-ray data of the samples was not found. One should also note that the morphology is very similar to pure ET samples (cp. Fig. 6.9)

phase (cp. Fig. 6.19). As an additional feature one should note the broad peak at approximately $2\theta = 14.8°$. This peak has been found in the co-evaporation measurements of ET and TCNQ. The origin is up to now unknown.

The measurement had a large background, suggesting that a large part of the sample is amorphous, and it was subtracted for clarity. This sample was prepared on a piece of a microscope slide, identical to those used in the test chamber. Therefore one can conclude that this type of substrate allows for growth of (ET)(TCNQ) and the growth of ET rather than (ET)(TCNQ) on the substrates prepared in the test chamber (cp. previous paragraph) has to be explained otherwise.

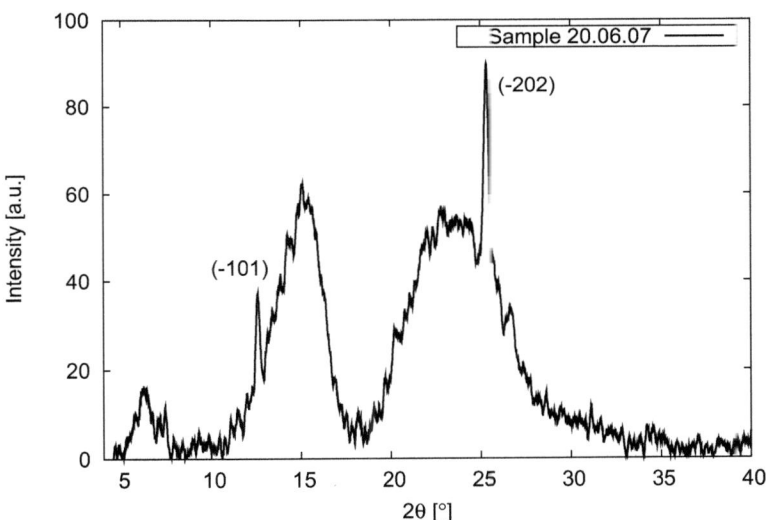

Figure 6.19: X-ray diffraction of sample 20.06.07
The first and second order of the monoclinic phase of (ET)(TCNQ) ($\bar{1}0l$) is observed. Additionally, there are two broad peaks at around $2\theta = 15°$ and $2\theta = 24°$. The peak at 15° might again be attributed to the sublimation of ET as seen in numerous other samples.

For two more samples, B8 and B11, X-ray diffraction provided results and ET ($0ll$) was identified. It should be noted that these two samples were prepared at elevated substrate temperatures, namely 100 °C and 140 °C, respectively.

For all other samples, no peaks were observed, so the phase of the crystallites could not be determined (cp. Table C.3).

OMBD - single crystals + TCNQ

The sublimation temperature found for (ET)(TCNQ) was the same as for pure ET. This suggests that the CT salt is decomposed to ET and TCNQ during sublimation. Since the sublimation temperature for pure TCNQ is lower than that of ET, this might lead to a deficiency

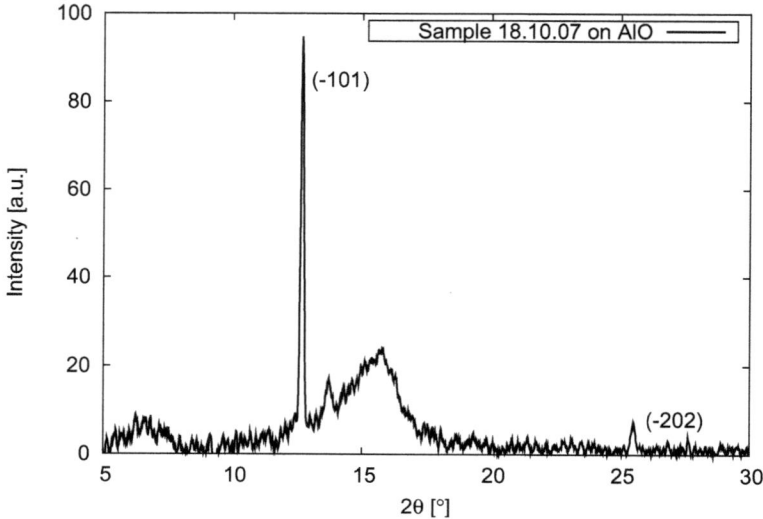

Figure 6.20: X-ray diffraction of Sample 18.10.07
The first and second order of the monoclinic phase of (ET)(TCNQ) $(\bar{1}0l)$ is observed. Additionally, there is a broad peak at around $2\theta = 15.4°$. This peak is a recurring feature of sublimation experiments including ET.

of TCNQ during preparation. Therefore the influence of additional TCNQ evaporation was studied.

Additionally, now usually two samples were prepared simultaneously on different substrates, Al_2O_3 and a Si substrate with a 300 nm SiO_2-layer on top. This will in the following be simply called SiO_2. These samples could then be used to determine the sulfur to nitrogen ratio with EDX. The method cannot be applied to samples prepared on Al_2O_3 substrates, since the substrate will charge due to the electron beam hitting the surface and will in turn deflect the beam.

For some samples pre-patterned SiO_2 substrates were used, meaning that aluminum pads had been prepared with sputtering and lithography on the substrate. For details of the preparation of the pre-patterned substrates, refer to [65, 66]. If a crystallite grows from one pad to another an electrical measurement could be carried out after preparation. An overview of the prepared samples is given in Appendix C, Table C.4.

For samples prepared on either Al_2O_3 a-plane or SiO_2 substrates (ET)(TCNQ) monoclinic phase could be observed in the $(\bar{1}0l)$ orientation. Sample C6 shall be pointed out, because here X-ray diffraction showed the monoclinic phase and simultaneously the EDX measurement confirmed a 2:1 ratio of sulfur to nitrogen. The diffraction pattern is shown in Fig. 6.20. Apart from the first and second order of the $(\bar{1}0l)$ orientation, again a broad peak at around 15.4° is present.

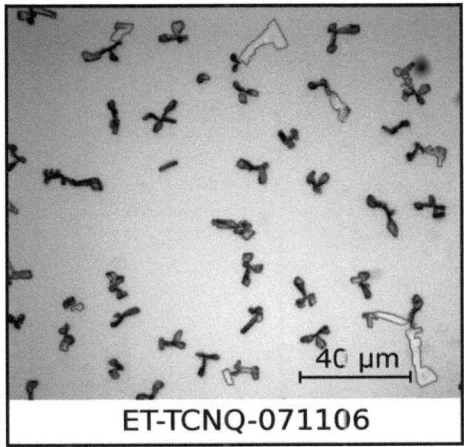

Figure 6.21: Optical micrograph of the sample from 06.11.07
Compared to the morphology of other (ET)(TCNQ) samples as shown in Fig. 6.18 no significant difference is observed.

Sample C9, prepared 06.11.07, was a surprise. An optical micrograph of the sample is shown in Fig. 6.21. Optically, the sample looks similar to the previously prepared ones, but it showed sharp peaks in X-ray diffraction (cp. Fig. 6.22), which could be identified as ET $(0ll)$. The peaks are visible up to fourth order. The rocking curves have a FWHM of 0.03° for (011), 0.04° for (022) and 0.05° for (033). Additionally, an EDX measurement was performed and only sulfur could be detected. The sample had been prepared at a substrate temperature of 100 °C. From these results the following conclusions can be drawn:

- Either ET or (ET)(TCNQ) can be prepared from (ET)(TCNQ) crystals and

- furthermore, ET can be prepared as highly ordered crystallites.

This strengthens the assumption from the pure ET experiments that impurities and ET fragments prevent the growth of an ordered thin film.

Conclusions

Independent of the additional evaporation of TCNQ from a second effusion cell (ET)(TCNQ) and ET could be deposited from single crystals of (ET)(TCNQ) on various substrates in the OMBD chamber. The morphology is always black, shiny microcrystallites and does not allow for the distinction between ET and (ET)(TCNQ) just from optical microscopy. The crucial parameter for the formation of the two observed phases seems to be substrate temperature. On substrates kept at 30 °C (ET)(TCNQ) monoclinic phase is observed in a $(\bar{1}0l)$ orientation. The

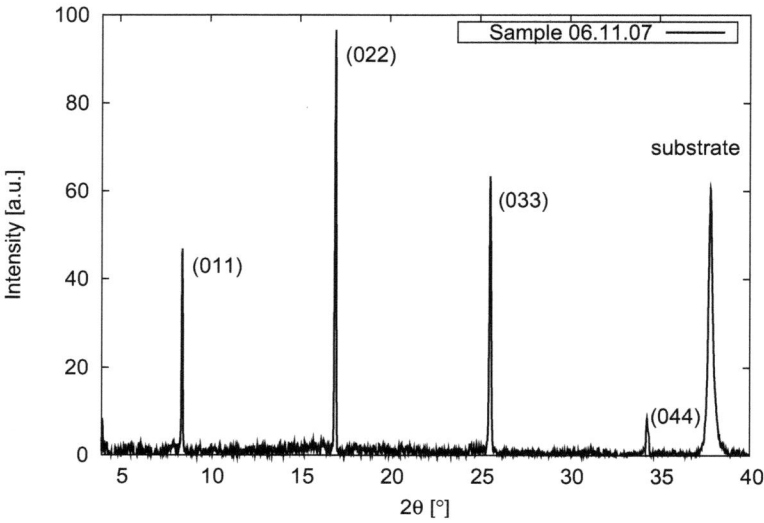

Figure 6.22: X-ray diffraction of sample 06.11.07
The X-ray diffraction shows sharp peaks of ET in the $(0ll)$ orientation. The peaks can be observed up to the fourth order and were visible after one scan in contrast to the previous samples, where numerous summation of repeated scans were necessary to reveal the peaks of (ET)(TCNQ). This shows that this is a highly ordered sample.

best results were obtained on Al_2O_3 a-plane, but also on Al_2O_3 random and SiO_2 the phase could be observed.

Above 100 °C ET in the $(0ll)$ orientation was found. This hypothesis is supported by the EDX data, which showed S and N for samples prepared at 30 °C and only S for samples prepared at higher temperatures, namely 60 °C, 100 °C and 140 °C.

One questions is, why no (ET)(TCNQ) was detected on the samples prepared by evaporation from (ET)(TCNQ) single crystals in the test chamber. An explanation would be an elevated substrate temperature, possibly caused by the effusion cell. With this assumption the data is consistent with the data obtained from deposition in the OMBD chamber. At the moment the substrate temperature can not be measured in the test chamber and is therefore unknown. Further influences could be the deposition rate and the cleanliness of the substrates, since the large microscope slides are much harder to clean in contrast to the 10×10 mm²-samples used in the OMBD chamber. Additionally, there is no bake-out function available for the substrates in the test chamber. For deposition in the OMBD chamber, samples are always baked out at 180 °C-400 °C, depending on the substrate material.

Further questions are the nature of the broad peak which has been observed in numerous samples. The central position shifts between $2\theta = 13° - 15.5°$ and also the width varies. This makes it difficult to determine, whether these peaks have really the same cause.

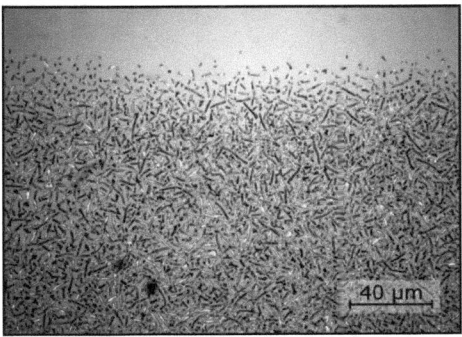

Figure 6.23: Optical micrograph of ET$_2$Cu(NCS)$_2$ - Sample 1
The morphology of the sample consists mostly of shiny, elongated crystallites. This reminds of the (ET)(TCNQ) samples prepared on microscope slides in the test chamber, where similar crystallites were found alternating with a black meander structure (cp. Fig. 6.16).

Summing up, the experiments with ET alone have shown that evaporation occurs close to the point of decomposition and that the resulting fragments or impurities prevent ordered thin film growth. The same is probably true for the formation of the CT salt (ET)(TCNQ) and could explain the difficulties to prepare the CT salt under co-evaporation of ET and TCNQ. Nevertheless, the data obtained by co-evaporation is currently not sufficient to draw a final conclusion whether a formation of (ET)(TCNQ) is possible or not with this method.

Evaporating from crystals of the preformed charge transfer salt, either ET or (ET)(TCNQ) growth was observed. Since the substrate temperature was typically lower than the evaporation temperature, this suggests, that the CT salt is broken up under evaporation, leaving the ET molecule intact, and that it is newly formed on the substrate, depending on the substrate temperature.

6.2.5 ET$_2$Cu(NCS)$_2$

Source material obtained from the reflux crystal growth experiment was mixed with quartz sand and put into a glass crucible. The sublimation experiments were carried out in the test chamber using the mini cell. The temperature of the cell was successively increased. Starting at 160 °C a deposition on the microscope slide was observed. The process was continued at 170 °C until a sufficiently thick deposition had occurred. This experiment could be repeated two times with a decreasing deposition rate. In the third experiment the temperature was successively increased to 200 °C, nevertheless the deposition had stopped. The crucible was thereafter checked and a fourth experiment was performed with temperatures up to 240 °C. No further deposition was observed.

From XRD (cp. Fig. 6.24) and optical microscopy it is concluded that ET was deposited from the ET$_2$Cu(NCS)$_2$ crystals. The morphology of the samples (cp. Fig. 6.23) is very similar to

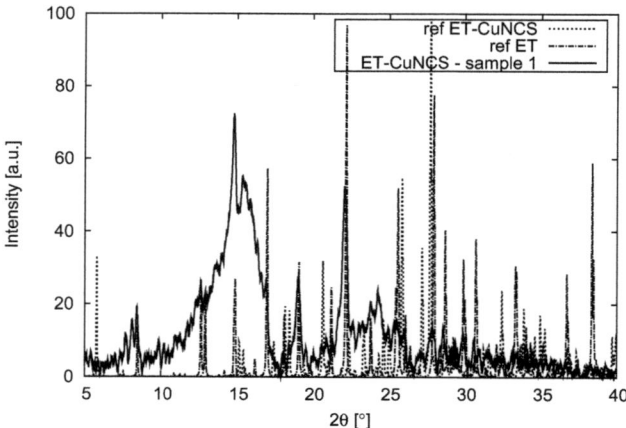

Figure 6.24: X-ray diffraction of $ET_2Cu(NCS)_2$ - Sample 1
In black the diffraction pattern of Sample 1 is shown. As comparison the reference data of ET is displayed in green, and the reference data of $ET_2Cu(NCS)_2$ in red. The measurement agrees well with the powder data of ET. In general the result is very similar as for the (ET)(TCNQ) samples prepared in the test chamber (cp. Fig. 6.17).

the morphology observed for the (ET)(TCNQ) evaporation in the test chamber (cp. Fig. 6.16). Furthermore, all observed peaks in XRD fit the reference of ET and the characteristic hump at about $2\theta = 15°$ is visible.

Chapter 7

Electronic Properties

To study the electronic properties of the various materials two types of measurements were performed: R(T) measurements and tunnel measurements.

7.1 R(T) measurements

7.1.1 (ET)(TCNQ) single crystals

A 2-probe resistance measurement on single crystals of the (ET)(TCNQ) β'-phase was performed. A crystal was glued on an Al_2O_3-substrate, which had been structured with 50 nm thick gold pads. As conductive glue silver paint was used. An optical micrograph of the crystal glued to the substrate is shown in Figure 7.1. Copper leads were attached to the gold pads to measure the resistance. The substrate was mounted on a Peltier element, to vary the temperature. The range between 300 K and 350 K could be measured with this setup. The reported metal-insulator transition at 330 K [37] for (ET) TCNQ β'-phase could be observed as a maximum in conductance/minimum in resistivity in the $I(T)$ measurement at constant voltage $U = 1$ mV. The experiment was performed in air and a degradation of the crystal could be observed. Several cycles of heating and cooling have been measured, and the resistivity of the crystal increased with every heating cycle. To demonstrate this behavior a full cycle is shown in Fig. 7.2. One can see that the measured current and accordingly the conductance $\sigma = \frac{I}{U}$ has decreased significantly after the first heating.

7.1.2 (ET)(TCNQ) thin films

Since the prepared thin films of (ET)(TCNQ) were generally microcrystalline, it was attempted to measure the resistivity of these crystallites by using either pre-patterned substrates or by evaporating metallic contacts over the prepared thin film.

Additional gold evaporation

The evaporation of metallic gold contacts using the stencil masks over an already prepared sample was tested with sample C1 (cp. Appendix C). A change of the micro-crystallites was

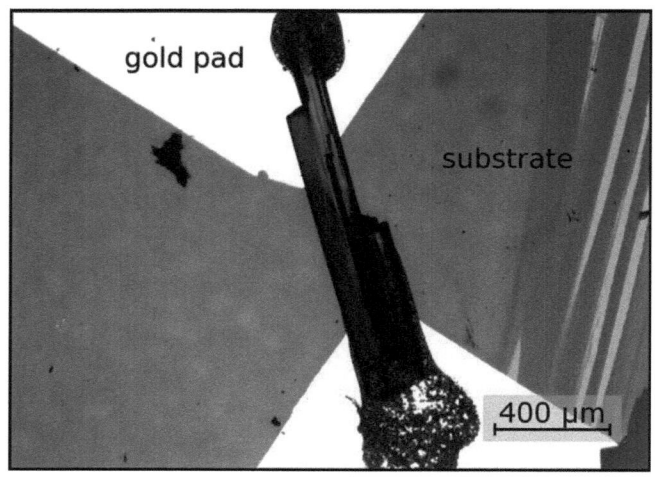

Figure 7.1: Optical micrograph of (ET)(TCNQ) crystal
An Al$_2$O$_3$-substrate was prepared with 50 nm gold pads. One crystal was glued over two pads with silver paint. The substrate appears dark, the gold pads appear white due to reflectance. On the gold pad in the lower right corner of the picture, the silver paint can be seen as a black-and white-dotted droplet.

observed (cp. Fig. 7.3). This might stem from the relatively high evaporation temperature of gold compared to the evaporation temperature of the organics. It is therefore expected that sample holder and substrate are heated significantly during the gold evaporation. For this reason it is suspected that the crystals of (ET)(TCNQ) may get damaged during the application of the gold pads. Nevertheless, 4 crystals were identified which crossed the gap between the two gold pads and an IV-curve of the sample was measured. Since the width of the crystals was about $(0.5 - 0.7)$ µm, the cross section of the crystals was estimated to $A = 4 \cdot \left(5 \cdot 10^{-7} \text{ m}\right)^2 = 10^{-12} \text{ m}^2 = 10^{-8} \text{ cm}^2$. With a current of $6 \cdot 10^{-10}$ A at 10 V the resistance is $R = 1.7 \cdot 10^{10}$ Ω. The gap width is about 17 µm and this corresponds to a resistivity of $\rho = \frac{RA}{l} = 9.8 \cdot 10^5$ Ωcm. This result is in good agreement with the resistivity of the monoclinic phase, where $\rho = 10^6$ Ωcm has been reported. Nevertheless, it has to be taken into account that the size of the crystallites could only be estimated very roughly, resulting in a large error of the resistivity value.

Pre-patterned substrates

SiO$_2$ substrates with aluminum contact pads were used as pre-patterned substrates for the evaporation. The aluminum had been sputtered onto the substrate and patterned by photolithography. For details on the preparation of these substrates compare [66]. With this method several problems occurred. First of all, the surface of the SiO$_2$ substrate was altered during the photolithography process. Regions, which differed in color, could be identified with the optical microscope and this inhomogeneity was also reflected in the growth of (ET)(TCNQ)

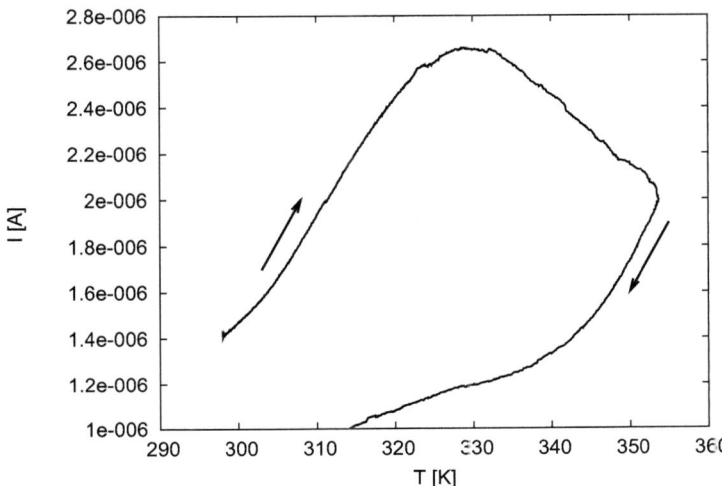

Figure 7.2: I(T) measurement
A constant voltage of 1 mV was applied to the crystal (cp. Fig. 7.1). The crystal was heated with the Peltier element from 300 K to 355 K and cooled back to 300 K. The course of time is indicated by the arrows. One can see a maximum in $I(T)$ at 330 K, which corresponds to a minimum in the resistivity.

and ET on the surface. This is visible in the left picture of Fig. 7.4, where a variation in density and size of the crystallites on the SiO_2 surface is apparent. Secondly, it turned out that the (ET)(TCNQ) crystallites avoid the gap between the aluminum pads. This effect is caused by a different mobility of the molecules on the SiO_2 surface compared to the aluminum surface. An indication for this is also the observance of very different density and size of crystals on the two different surfaces. This effect can also be observed in the left picture of Fig. 7.4. On the right hand side of Fig. 7.4 a detailed view of the contact area, designated for the crystal growth, is shown. The distance between two adjacent aluminum pads is 5 μm. Although a clustering of the crystallites at the step edge of the aluminum is seen and despite the fact, that crystals with a length of 5 μm were easily observed on non-structured substrates, we did not succeed to grow a crystal over the gap of the pads and measure the conductivity.

7.2 Tunnel Measurements

Tunneling spectroscopy is a technique used to study superconductors. It was first proposed 1960 by Giaever [10, 11]. Information about the order parameter can be obtained and the density of states can directly be measured. Although most powerful for superconductors, tunneling spectroscopy has also been applied to semiconductors. It has been shown, that the band structure of semiconductors can be studied, but most work has been done for highly degenerated semiconductors and semimetals [67, 68, 69].

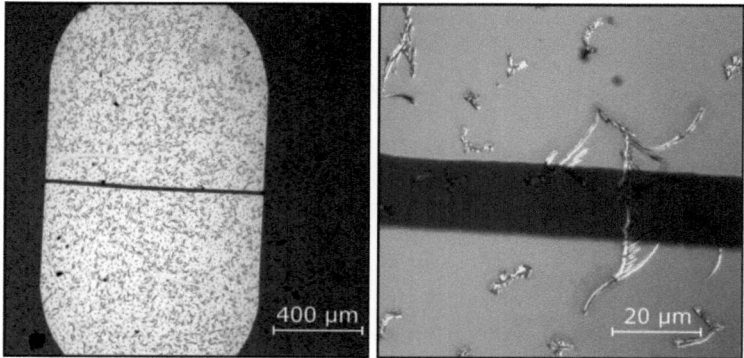

Figure 7.3: Optical micrographs of sample C1

The pictures show the (ET)(TCNQ) sample C1 after an additional deposition of gold pads onto the sample. On the left hand side, the two pads are nearly completely visible, whereas on the right hand side only a part of the gap is shown with higher resolution. A crystal reaching over the gap is observed. Additionally one can see a darker color around crystals that lie under the gold pads. This is interpreted as a reaction of the crystal with the gold.

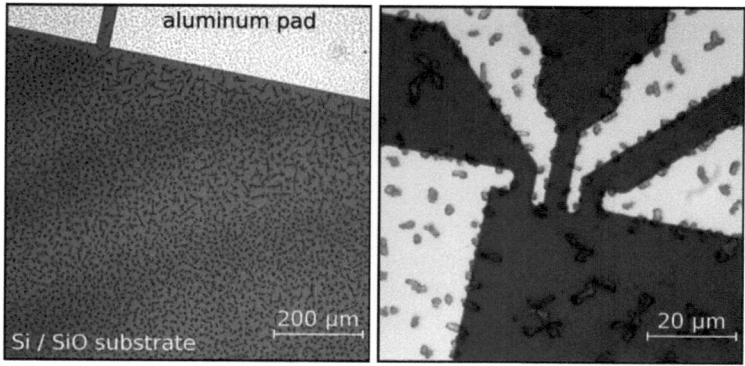

Figure 7.4: Optical micrographs of sample C10

The pictures show the (ET)(TCNQ) sample C10 on the Si/SiO$_2$ substrate. The aluminum pads appear in light color, the surface of the Si/SiO$_2$ substrate appears dark. On the left hand side the inhomogeneous growth on the substrate and the aluminum pads can be seen. On the right hand side a detailed view of the designated contact area is seen. The crystallites cluster at the edges of the aluminum pads, but do not grow over the gap between adjacent pads.

7.2.1 Measurement principle

Two electrodes A and B are separated by a thin insulating barrier, the so-called tunneling barrier. In the case of a superconductor-insulator-normal metal junction (SIN) it has been shown, that dI/dV is directly proportional to the density of states of the superconducting electrode. This is often visualized in the semiconductor picture, shown in Fig. 7.5.

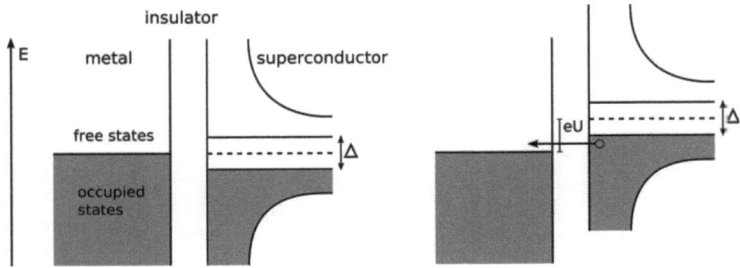

Figure 7.5: SIN junction in the semiconductor picture
On the left hand side a normal metal-insulator-superconductor junction is depicted with the energy as y-axis and the density of states (DOS) as x-axis. The right hand side shows the situation if an external voltage U is applied. Electrons may tunnel from occupied states of the superconductor to free states of the metal. The number of electrons - and therefore the current - is directly proportional to the DOS of the superconductor.

7.2.2 Planar tunnel junction

In the conventional cross junction geometry the two electrodes A and B are prepared as thin film stripes, where the second electrode is prepared orthogonal to the first one, so that they form a cross. A typical preparation sequence is shown in Fig. 7.6. The insulating barrier has to be prepared before applying the second electrode, which is often done by oxidizing the first (metal) electrode to form a natural oxide. This allows for an approximate four terminal

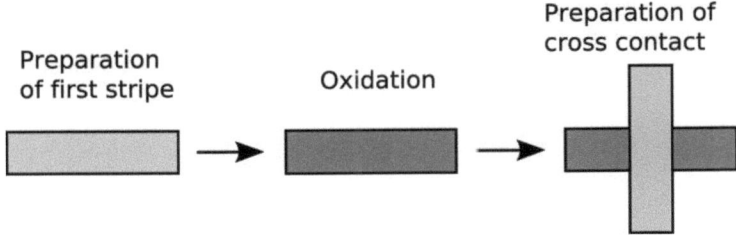

Figure 7.6: Planar cross junction geometry
The typical preparation steps for a cross junction are preparation of a metallic stripe, then oxidation to form the tunnel barrier and finally the preparation of the cross contact.

measurement, if the resistance of the tunnel contact is much larger, than the resistance of the electrodes.

Organic semiconductors have a high resistivity (10^6 Ωcm) which prohibits to use the conventional planar tunnel contact structure, because the organic electrode adds a large resistance in series to the tunnel junction. To overcome the problem of the poorly conducting semiconductor electrode an additional metal layer is introduced as shown in Fig. 7.7. This reduces the total resistance of the device and ensures that the main voltage drop occurs over the tunnel barrier. As a drawback this also means, that one introduces an additional metal-semiconductor interface to the device. This may in most cases lead to a Schottky contact in series to the tunnel contact.

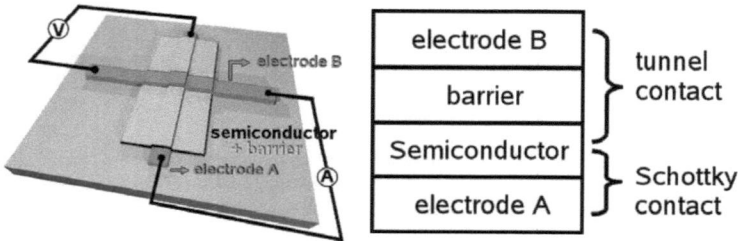

Figure 7.7: Modified junction geometry for semiconductors
On the left hand side a 3D model of the new geometry is shown, on the right hand side the horizontal stacking of the different layers in the contact area is shown.

7.2.3 Measurement setup

The principle measurement setup is shown in Fig. 7.8. A current-adding circuit is connected to two terminals of the tunnel contact. At the two remaining terminals the voltage drop over the contact is measured.

A Lock-In amplifier provides a small AC-current $I \sin(\omega t)$ which is added to the DC current I_0. This modulated current is then applied to the tunnel junction. The AC voltage drop over the junction is detected with the Lock-In and the DC voltage with a nanovoltmeter. The measured AC voltage is directly proportional to $\frac{\partial U}{\partial I}$ according to

$$U(I_0 + I \sin(\omega t)) = U(I_0) + \left.\frac{\partial U}{\partial I}\right|_{I_0} I \sin(\omega t) + \frac{1}{2} \left.\frac{\partial^2 U}{\partial I^2}\right|_{I_0} I^2 \sin^2(\omega t) + ... \quad (7.1)$$

By sweeping the applied DC current I_0 a so-called tunnel spectrum (dI/dU over U) is measured.

7.2.4 Device preparation

Several tunnel junctions were prepared with different electrode materials. An overview is given in Table 7.1. A picture of a complete device is shown in Fig. 7.9. One sample consist always of

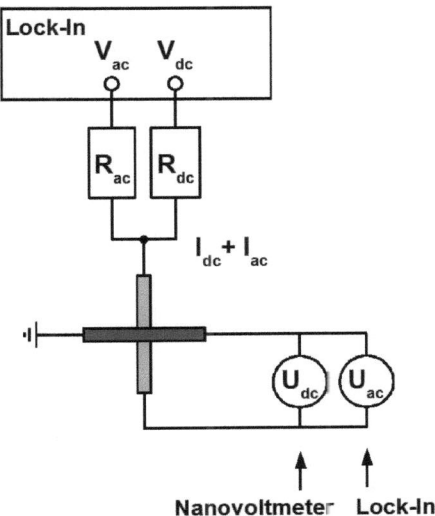

Figure 7.8: Tunneling measurement setup
AC and DC output of the Lock-In (V_{ac}, V_{dc}) are coupled over two resistors to form a current adding circuit. This is connected to two terminals of the tunnel contact and at the other two terminals the resulting voltages U_{dc} and U_{ac} are measured separately with a nanovoltmeter and the Lock-In.

four tunnel contacts prepared simultaneously on one substrate. All devices were prepared on Al_2O_3 substrates. The principal preparation steps were very similar for all samples. First an aluminum stripe was sputtered through a stencil mask and then oxidized. The details on the aluminum/aluminum oxide preparation can be found in Appendix D. The sputtering causes blurry edges of the aluminum stripe due to the following facts:

- the distance sample/target is small (ca. 2 cm),
- the target diameter is comparatively large 2 inch and
- the sputtered atoms collide with gas atoms and impinge therefore with random directions onto the sample.

For T1-T4 a metallic counter electrode was prepared after oxidation. For T5-T7 CuPc was evaporated through a stencil mask in the OMBD chamber and finally, a metallic counter electrode was deposited. In general the following conclusions can be drawn:

- Indium was not suitable as a counter electrode, since it oxidized too fast.
- Reactive sputtering (in an Argon/Oxygen atmosphere) of AlO_x did not provide a decent tunnel barrier.

No.	Structure	Comment	Tunnel experiment
T1	Al-AlO$_x$-In	Indium oxidized	unsuccessful
T2	Al-AlO$_x$-Nb	prepared with sample rotation	see Fig. 7.10
T3	Al-AlO$_x$-Nb	prepared by reactive sputtering	unsuccessful
T4	Al-AlO$_x$-Nb	sample broke	unsuccessful
T5	Al-AlO$_x$-CuPc-Au	-	-
T6	Al-AlO$_x$-CuPc-Nb	-	see Fig. 7.11
T7	Al-AlO$_x$-CuPc-Au	-	unsuccessful

Table 7.1: Overview over the prepared tunnel contacts

7.2.5 Results

The measurements were carried out in a He^4 cryostat. The setup of the sample insert can be found elsewhere [70]. The Al-AlO$_x$-Nb-contact (T2) served as a test sample to study the measurement setup with a conventional SIN structure before measuring a junction with a new layer structure and new materials. The superconducting gap of the Niobium electrode was measured, which shows a successful device preparation and a functional measurement setup. The result of the measurement is shown in Fig. 7.10. The spectrum is smeared out, since the sample was prepared using a rotating sample holder. The Nb electrode had a gradual thickness distribution leading to a successive inset of superconductivity for the different regions.

As second successful experiment sample T6, an Al-AlO$_x$-CuPc-Nb structure, was measured. The tunnel spectrum is shown in Fig. 7.11. The measurement shows a parabolic background conductivity, which is in agreement with the Simmons model [71]. Around zero bias voltage an additional feature is visible, namely, at around 60 mV an additional conduction channel opens up. The slight asymmetry of the feature is attributed to a Schottky contact of the CuPc to the Niobium. Energetically, a phononic excitation is suspected. The corresponding wave number is approximately 480 cm^{-1}. To identify the respective excitation the IR spectrum of the molecule was examined[1] and a broad peak at around 450 cm^{-1} can be seen. The nature of this excitation can not easily be determined, because it lies already within the fingerprint region of the molecule. This means that it is generally not possible to assign the signal to a single functional group, since the fingerprint region mainly contains bands from interacting vibrational modes. Stretching modes of C-H, C-N, N-H or C=C can be ruled out, since the corresponding wave numbers lie between 1500 − 3500 cm^{-1}. Since CuPc was only chosen for this experiment because of its robustness, but not because of its electronic properties, this was not studied further.

This first experiment with the new layer structure serves as a proof of concept. It shows on the one hand that the device structure is functional, but more importantly, that features in the tunnel spectrum can be expected for organic materials.

---

[1] cp. spectral database: http://riodb01.ibase.aist.go.jp/sdbs/cgi-bin/cre_index.cgi

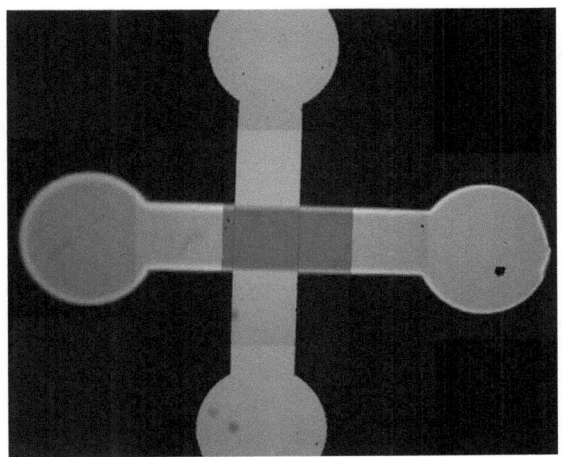

Figure 7.9: Optical micrograph of Al-AlO$_x$-CuPc-Au tunnel junction (T5)
The picture has been merged from several smaller photographs showing only a part of the structure. The structure has been prepared in the following order: First the horizontal aluminum bar was prepared by sputtering, causing the rather blurry edges. It was then oxidized. Afterwards the CuPc rectangle was prepared by thermal evaporation and at last the vertical gold stripe was also prepared by thermal evaporation. The active area (region where the stripes cross) is 0.5×0.5 mm^2.

However, the interpretation of such measurements demand detailed examination of the electronic structure of the organic thin film and the device as such.

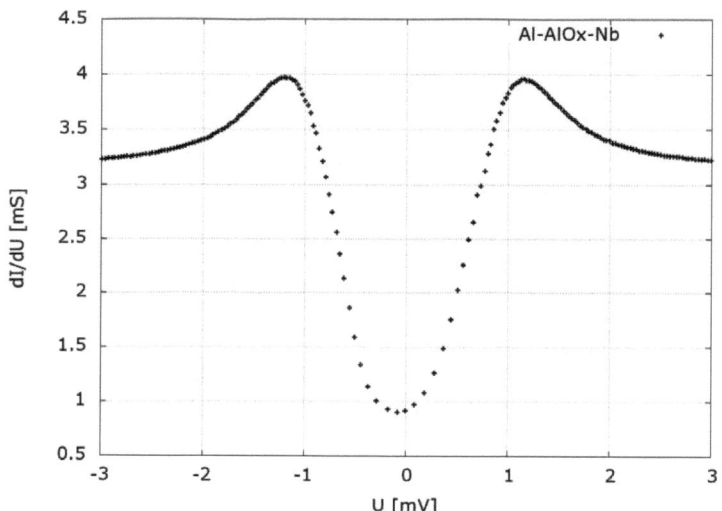

Figure 7.10: Tunnel spectrum of an Al-AlOx-Nb contact
The spectrum was measured with a modulation frequency of $f = 117$ Hz, resistors of $R_{ac} = R_{dc} = 1$ MΩ an modulation amplitude of 0.3 V at a temperature of 1.7 K. The spectrum is somewhat smeared out, because the Nb electrode had a inhomogeneous thickness.

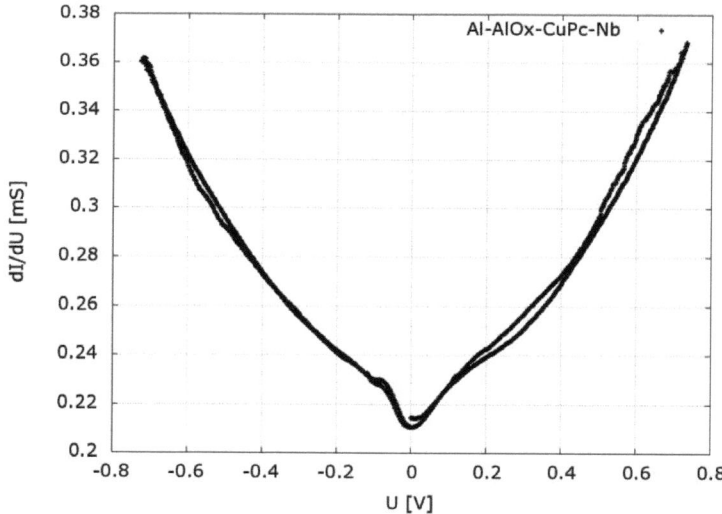

Figure 7.11: Tunnel spectrum of an Al-AlOx-CuPc-Nb contact
The spectrum is an average of 5 measurements. The spectrum was measured with a modulation frequency of $f = 117$ Hz, resistors of $R_{ac} = R_{dc} = 1$ MΩ an modulation amplitude of 5 V at a temperature of 2 K. The Nb gap, with a width of approx. 1 mV, can here in contrast to Fig. 7.10 not be resolved.

Chapter 8

Summary and Outlook

During this thesis an organic molecular beam deposition system was designed and successfully integrated in the existing MBE setup. The setup was furthermore extended with a load-lock, a chamber for contact preparation and a sputter chamber. The whole system offers the possibility to prepare thin films with electron beam evaporation, sputtering and thermal evaporation with effusion cells. From a material point of view organics, metals and dielectrics are available.

A stencil mask system has been developed to enable in-situ device preparation. Structures down to 20 µm have already been prepared and structures down to a size of 12.5 µm are available.

This setup is instrumental basis for the preparation of organic thin films, and furthermore, for thin film devices such as tunnel contacts or field effect transistors.

With a preliminary setup, the final MBE system and a test chamber the preparation conditions for a variety of organic molecules and charge transfer complexes have been explored, where the preparation of $Cu(NCS)_2$ and $ET_2Cu(NCS)_2$ has not been successful. A focus was set on ET, TCNQ and the CT salt which is formed by the two, (ET)(TCNQ). For (ET)(TCNQ) conditions for the growth of the monoclinic phase have been found.

The main characterization methods for the organic thin films were optical microscopy and X-ray diffraction. The optical microscopy showed that the samples usually exhibit microcrystalline structure with various different features. It was therefore impossible to categorize the samples from this point of view. Due to the micro-crystallinity and the resulting high surface roughness, X-ray reflectometry could not be applied successfully. With X-ray diffraction several phases could be identified, but first after the typical measurement times had been extended to several hours for a single sample!

For the organic materials one crucial task for the future is to improve growth towards a closed layer. The purity of the source material has been identified as an important parameter for the film quality and it is therefore recommended to employ methods for purification of the source material prior to evaporation.

Since several of the studied materials were not suited for sublimation, a search for interesting organic materials with better sublimation and growth behavior should accompany further experiments. The test chamber can serve as a tool for screening new materials with respect to their evaporation parameters.

First tunnel contacts using organics have been prepared to establish the device preparation process including the stencil masks and to verify the measurement setup for device characterization.

Parallel to the growth experiments a simulation program has been developed and tested to predict evaporation characteristics and find optimum cell geometries. The program hereby includes all main processes known to influence the evaporation characteristics, namely, evaporation from source material and crucible walls, migration on the crucible walls, collisions between molecules in the gas phase and the effect of temperature gradients within the crucible. Several types of crucibles have been simulated with the aim to identify an optimized geometry for the evaporation of organics, which evaporate close to the point of decomposition with relatively low evaporation rates.

As an experimental verification different simulated effusion cells were realized and their evaporation characteristics were studied. Here, the problems were mainly found on the experimental part, where the thickness measurement of the prepared samples proved to be difficult. So far, the simulation program has provided results in good agreement with the experimental work presented here, as well as with experimental and theoretical work of other groups.

For the simulation program, only a small set of parameters has been studied so far. This leaves room for many questions to be tackled. Up to now mainly single processes (only evaporation, only migration) have been studied or a combination of two. The competition of all possible processes within one simulation is still to be examined.

With the application of the simulation to EBID geometries it has already been shown that the program is quite universal. Both the simulated geometries as well as the models for the implemented processes can be advanced and possibly also specialized to specific applications. Hereby, it can be studied up to which point approximations and simplifications are valid, and where advancements really influence the simulation results. An important task here is to simultaneously develop new experiments to verify or disprove the simulation results.

Appendix A

UHV components

A.1 OMBD chamber

No.	size	function	position/comment
6	CF63	effusion cells	500 mm from the sample plane
1	CF40	QMB	visible from all effusion cell ports
1	CF16	QMB shutter	-
1	CF150	sample manipulator	-
1	CF16	sample shutter	top
1	CF40	cooling mechanism	-
1	CF40	pressure gauge	-
1	CF16	pressure gauge	-
1	CF40	RHEED cathode	tilted 3°
1	CF150	RHEED screen	opposing the RHEED cathode
1	CF100	mass spectrometer	-
2	CF40	reflectometry	tilted 45°, opposing each other
1	CF150	LEED optics	tilted 20° to sample plane
1	CF100	CMA*	tilted 45° to sample plane
1	CF40	pyrometer	bottom, along sample normal
1	CF200	pumping system	-
1	CF100	transfer connection	-
6	-	view/spare ports	

Table A.1: List of all ports of the OMBD chamber
The designated function and resulting positions and/or tilt angles are given. The tilt angle is measured relative to the sample plane. * cylindrical mirror analyzer

A.2 Stencil masks

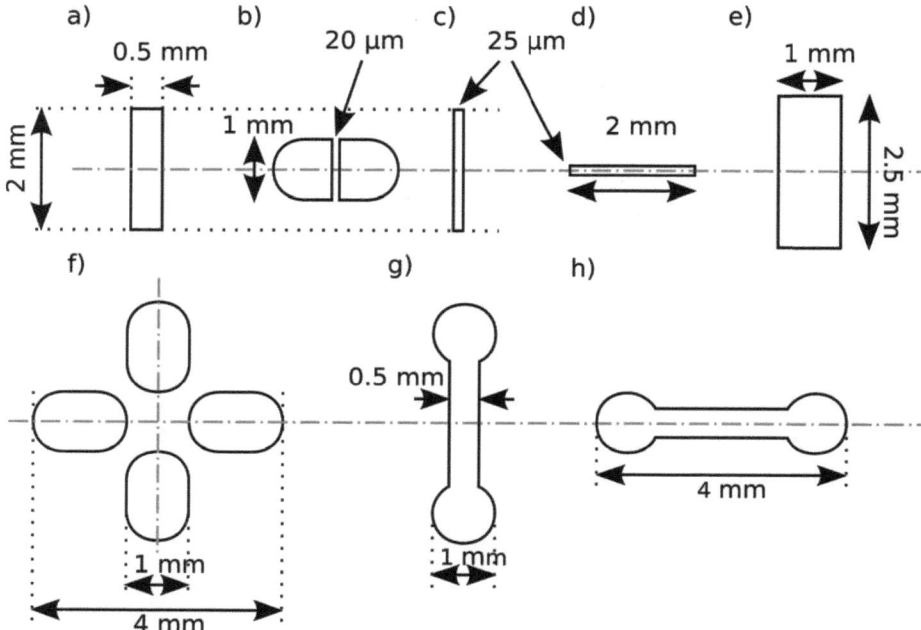

Figure A.1: Sketch of the currently available stencil mask types
Each structure is arranged four-fold on one stencil mask, so that four devices are prepared simultaneously on a 10×10 mm^2 substrate.
a) Gate contact and active area for field effect transistors,
b) drain and source contact for field effect transistors,
c) electrode A for tunnel contacts (narrow),
d) electrode B for tunnel contacts (narrow),
e) contact areas for tunnel contacts (to be used in combination with c) and d)),
f) dielectric for field effect transistors,
g) electrode A for tunnel contacts (large),
h) electrode B for tunnel contacts (large)

Appendix B

Additional simulations

Here additional simulations are presented, which demonstrate the versatility of the simulation program and comment on aspects that have not been covered exhaustively in chapter 5. The results presented here are of preliminary nature and need further investigation.

B.1 Migration and temperature gradient

Migration was studied in a simulation with no evaporation, but different temperature gradients. An open cell was simulated. The bottom of the cell acts as an infinite source as usual and the open end of the cell is treated as a sink of particles. This behavior stems from the fact that particles reaching the uppermost area element will be subtracted and "leave" the simulation, if an additional upwards movement takes place. With one source of particles, one sink of particles and a homogeneous temperature distribution one expects a linear gradient of particles along the cell. This is nicely confirmed by the simulation as shown in the left part of Figure B.1. The number of particles decreases linearly with height and is constant in ϕ.

As a next step a temperature gradient was included either increasing or decreasing linearly with height. The results are shown in the right part of Figure B.1. The lower temperature was hereby identical to the constant temperature in the first simulation. If $T_{top} > T_{bottom}$, the top area was depleted by particles, which can move more easily due to the higher temperature. On the contrary, if $T_{top} < T_{bottom}$ a pile-up of particles was produced, since more particles are delivered by the "hot" source, but cannot be transported fast enough towards the sink.

B.2 Temperature gradients

Temperature gradients have also been included in simulations with evaporation, but only a very weak dependence has been observed. Very helpful is the extreme case of 0 K for the walls, which can be used to separate the direct contribution of the source from the contribution of the cell walls. With this method the geometrical cut-off of the thickness distribution on the plane due to the crucible geometry can directly be observed.

It has up to now not been included that the geometry of the crucible is influenced, if a

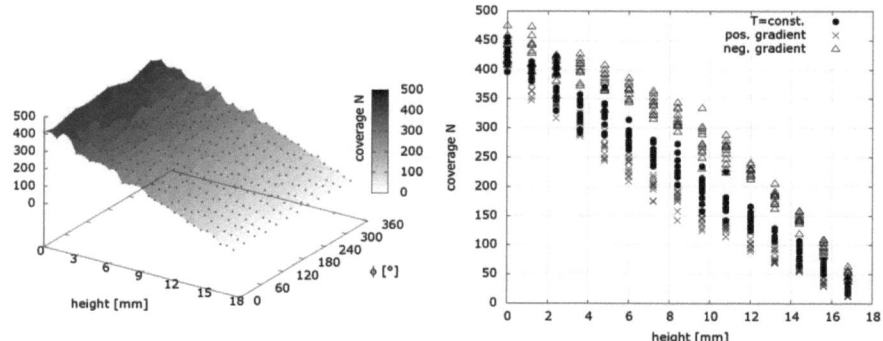

Figure B.1: Wall coverage of an open cell after migration
An open cell of $r = 3$ mm and $L = 18$ mm was simulated. On the left hand side a 3D view of the wall coverage for a cell with homogeneous temperature is shown. On the right hand side a comparison of a cell with homogeneous temperature, positive gradient, $T_{\text{top}} > T_{\text{bottom}}$ and negative gradient, $T_{\text{top}} < T_{\text{bottom}}$ is shown. Since the gradient was applied along the height of the simulated cell, no angular dependency is present. Therefore the 3D data is presented as a projection onto the plane of coverage and height.

significant coverage is reached, namely that the cell eventually closes due to recrystallization of material at cold spots of the cell.

B.3 Tilted substrate

The program is capable of simulating tilted substrates, which is important for MBE, since the effusion cells are usually mounted tilted to the sample for the simple reason to give room for several effusion cells in a symmetric arrangement with respect to the sample. The problem of the asymmetric evaporation characteristics due to the tilt is sometimes counteracted with a sample rotation.

A second area of application is electron beam induced deposition (EBID) which is carried out in a scanning electron microscope (SEM). A capillary with effusing gas is brought close to a substrate and deposition occurs through bombardment with the electron beam, which dissociates the gas at the substrate. Experimental work on this topic can be found in [65, 66].

Here a tilted arrangement is absolutely necessary, because the electron gun is by default mounted top down and the capillary has to be inserted diagonally from the side. A sample rotation is usually not implemented and therefore it is even more important to investigate the characteristics of the capillary with respect to a tilted substrate. A typical EBID geometry is shown in Fig. B.2.

Straight capillary To represents a simple straight capillary an open cell of $r = 2$ mm and $L = 20$ mm was simulated. The typical size of the capillary is a factor of 10 smaller, but the results can be scaled. The plane distance was chosen to 6 mm. The results are shown

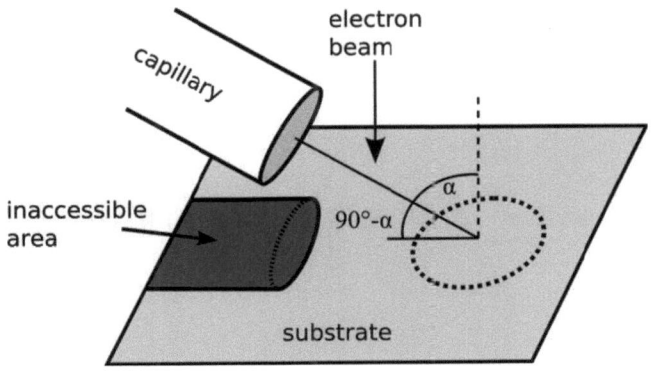

Figure B.2: Sketch of typical EBID geometry
A substrate is shown with a capillary approaching from the left and bombardment with the electron beam from above. The unaccessible area for the electron beam is indicated by the shadow of the capillary on the substrate. Additionally the projection of the capillary onto the substrate is indicated by the dotted circle.

in Fig. B.4. The capillary approaches the substrate from the negative x-direction and for clarity the projection of the capillary is included, to mark the area that is not accessible with the electron beam. In an optimum arrangement the electron beam can access the area of highest coverage. This is best achieved with a tilt angle α of 40°-60° in the simulation. In an experimental setup this is indeed a typical tilt angle of the capillary.

Utke et al. has also performed a Monte Carlo simulation for EBID for a tilt angle of 50° for a capillary of $r = 300$ μm and a plane distance of 614 μm [72]. To compare to his results a second simulation with $r = 3$ mm and otherwise identical parameters as before was carried out. The results are shown in Figure B.5. In contrast to the previous simulation the area of maximum coverage is now always nearly in the center of the projection of the capillary and therefore shaded from the electron beam. The result for 50° is very similar to Utke's result.

This shows that the ratio between diameter of the capillary and distance from the sample plays an important role for the thickness distribution on the sample.

Beveled capillary To uncover the area of maximum coverage to the electron beam a capillary with beveled top can be imagined. With the long edge pointing toward the substrate the accessible area could be increased (cp. Fig. B.3. To prove this hypothesis an according simulation was performed. The results are shown in Fig. B.6. As expected the area with highest coverage is indeed accessible, but another aspect has to be considered: For efficient material usage and deposition the *absolute* coverage is of importance. For a constant evaporation temperature and therefore constant flux of particles a maximum amount of particles at the position of the electron beam is desirable.

Therefore the simulations in Fig. B.5 and Fig. B.6 have been performed with identical pa-

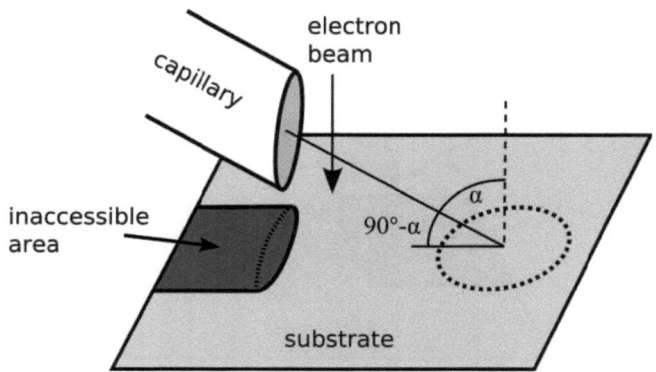

Figure B.3: Sketch of EBID geometry with beveled capillary
A substrate is shown with a capillary approaching from the left and bombardment with the electron beam from above. The unaccessible area for the electron beam is indicated by the shadow of the capillary on the substrate. Additionally the projection of the capillary onto the substrate is indicated by the dotted circle.

rameters except for the geometry of the capillary. Thus the absolute coverage on area elements accessible to the electron beam can directly be compared. As a result, the absolute coverage for the simple capillary is higher than for the beveled model.

In conclusion, implementing the tilted substrate the simulations can be extended to a new field of application, namely EBID. Here, a maximum coverage rather than a homogeneous distribution might be of interest, because of the small scale on which the electron beam accesses the sample area compared to the typical size of the capillary. For this field also a experimental verification of the results is desirable to prove the predictive power of the simulation program.

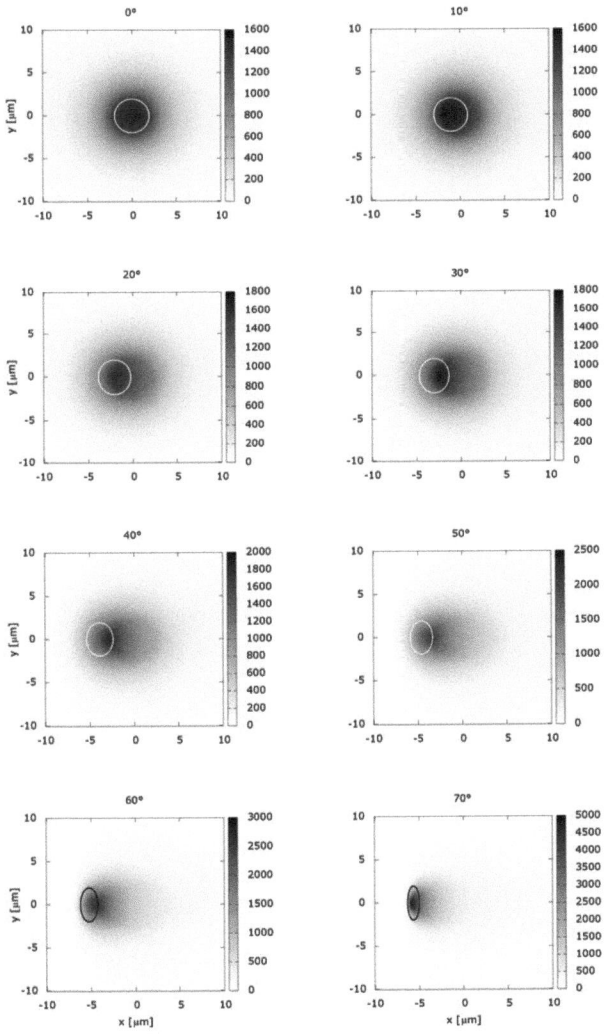

Figure B.4: Thickness distribution for tilted substrates
An open cell with $r = 2$ mm and $L = 20$ mm was simulated. The thickness distribution is shown on tilted substrates from 0° to 70° with a plane distance of 6 mm. The capillary is assumed to approach from the left of each graph and the ellipse indicates the projection of the capillary onto the substrate. The right edge of the ellipse therefore limits the area, which is accessible with the electron beam.

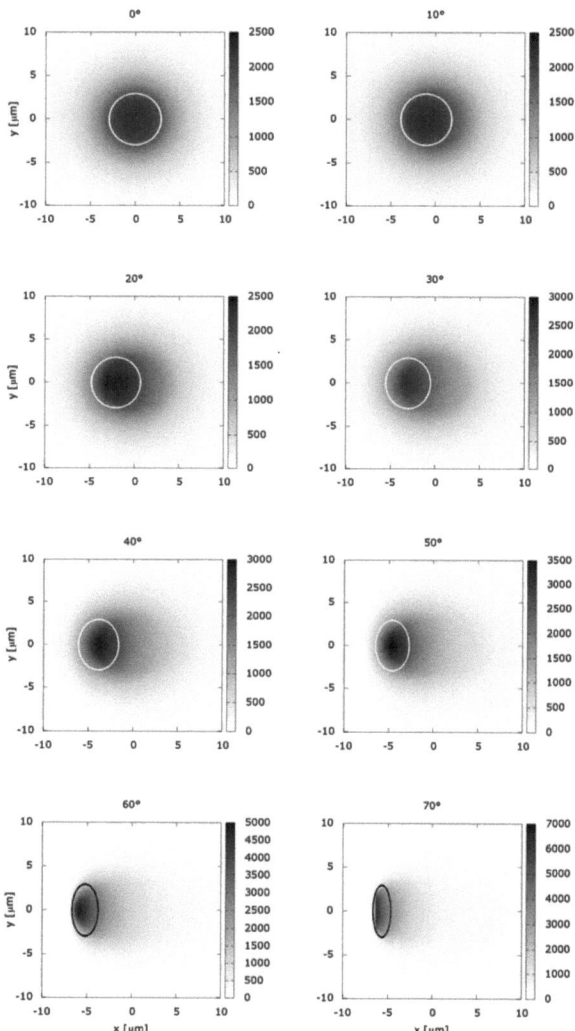

Figure B.5: Thickness distribution for tilted substrates
To compare to the results of Utke et al. an open cell with $r = 3$ mm and $L = 20$ mm was simulated. The thickness distribution is shown on tilted substrates from 0° to 70° with a plane distance of 6 mm. The capillary is assumed to approach from the left of each graph and the ellipse indicates the projection of the capillary onto the substrate. The right edge of the ellipse therefore limits the area, which is accessible with the electron beam.

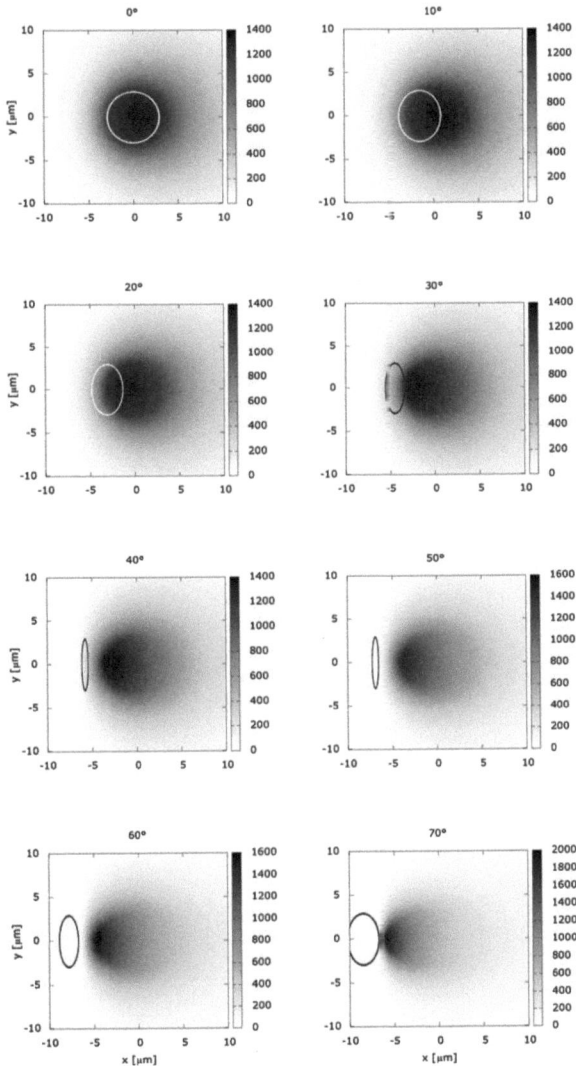

Figure B.6: Thickness distribution for a beveled capillary
In analogy to Fig. B.5 an open cell with $r = 3$ mm and $L = 20$ mm was simulated. The capillary has a beveled edge of 45°. The thickness distribution is shown on tilted substrates from 0° to 70° with a plane distance of 6 mm. The capillary is assumed to approach from the left of each graph and the ellipse indicates the projection of the capillary onto the substrate. The long edge of the capillary is pointing towards the substrate as depicted in Fig. B.3.

Appendix C

Sample Overview

No.	Date	T_{et} [°C]	T_{sub} [°C]	t [s]	t [h]	substrate
ET1	12.01.07	120	RT	11830	3.3	Al_2O_3 random
ET2	16.01.07	125	60.2	11450	3.2	Al_2O_3 random
ET3	23.01.07	127	RT	10110	2.8	Al_2O_3 random
ET4	26.01.07	130	RT	10800	3	Al_2O_3 random
ET5	29.01.07	133	RT	11760	3.3	Al_2O_3 random
ET6	31.01.07	133	RT	29100	8.1	glass
ET7	01.02.07	136	RT	10800	3.0	Al_2O_3 random
ET8	17.02.07	136	-14.7	9060	2.5	Al_2O_3 random
ET9	23.02.07	142	RT	8350	2.3	Al_2O_3 a-plane
ET10	26.02.07	148	RT	7200	2.0	$SrTiO_3$ (001)
ET11	26.02.07	154	RT	3800	1.1	$SrTiO_3$ (001)
ET12	27.02.07	148	RT	7200	2	Al_2O_3 random
ET13	02.03.07	150	RT	4300	1.2	**
ET14	06.03.07	158	RT	8800	2.4	Al_2O_3 *
ET15	21.03.07	140	RT	65200	18.1	glass
ET16	04.04.07	120-125	30	17150	4.8	Al_2O_3 a-plane
ET17	11.04.07	135	30	18280	5.1	Al_2O_3 random

Table C.1: ET evaporation from graphite and quartz crucibles
T_{et} denotes the temperature of the ET cell and T_{sub} the substrate temperature. t is the deposition time in seconds or hours. RT signifies room temperature and corresponds to $30-35$ °C due to the radiation heating from the cell. Samples ET1-ET15 were prepared from a graphite crucible, ET16 and ET17 from a glass crucible.
** Al_2O_3 random substrate onto which Au had been deposited prior to the experiment
* orientation unknown

Date	T_{tcnq} [°C]	t [h]	d [Å]	substrate
20.12.06	105	3	32	Al_2O_3-random
21.12.06	110	3	105	Al_2O_3-random
19.01.07	115	4.33	2400	Al_2O_3-random
22.01.07	115	0.75	210	Al_2O_3-random
05.04.07	115	2	290	Al_2O_3 a-plane
10.04.07	120	12	1936	Al_2O_3-random
06.06.07	120-125	9.8	1050	Al_2O_3 *
11.06.07	125	3.45	260	glass
12.06.07	120-130	7.25	803	glass

Table C.2: Overview of the prepared samples of TCNQ

T_{tcnq} and T_{sub} denote the temperature of the TCNQ cell and the substrate, respectively. t signifies the deposition time and d the thickness as displayed with the QMB. Between 22.01.07 and 05.04.07 the source material was exchanged and the new material was mixed with quartz sand. * orientation unknown

No.	Date	$T_{et-tcnq}$ [°C]	T_s [°C]	t [s]	t [h]	p [mbar]	XRD
B1	06.06.07	0-105	30	-	0	max-1E-6	/
B2	15.06.07	110-115	30	-	0	5.00E-8	-
B3	18.06.07	120	30	21800	6.06	3.30E-8	-
B4	19.06.07	125	30	21600	6	5.00E-8	-
B5	20.06.07	125	30	28000	7.78	4.80E-8	mono
B6	25.06.07	125	60	25300	7.03	2.70E-8	-
B7	27.06.07	125	80	25300	7.03	2.30E-8	-
B8	28.06.07	125	100	31150	8.65	2.10E-8	ET
B9	02.07.07	125	120	51700	14.36	1.50E-8	-
B10	03.07.07	125	RT?	50700	14.08	1.50E-8	/
B11	09.07.07	125	140	26800	7.44	1.00E-08	ET
B12	10.07.07	125	160	33900	9.42	1.10E-08	-
B13	11.07.07	125	140	33600	9.33	9.00E-09	/

Table C.3: Evaporation of (ET)(TCNQ)

The table shows the (ET)(TCNQ) samples, prepared by sublimation from single crystals. $T_{et-tcnq}$ is the cell temperature, T_s the substrate temperature, t the deposition time and p the pressure during deposition. The key for the result of the x-ray measurement (XRD) is the following: / = no x-ray measurement performed, - = no peaks, mono = (ET)(TCNQ) monoclinic phase, ET = ET

No.	Date 2007	T_{e-t} [°C]	T_t [°C]	T_s [°C]	t [s]	t [h]	d [kÅ]	p [mbar]	sub1	S:N	xrd1	sub2	xrd2
C1	17.07.	125	120	30	27200	7.56	0.29	7.00E-8	SiO_2	2:1	-	Al_2O_3 (r)	-
C2	23.07.	125	120	30	27000	7.5	0.27	6.60E-8	SiO_2	2:1	/	-	-
C3	25.07.	125	120	30	27000	7.5	0.25	8.00E-8	SiO_2	-	/	-	-
C4	25.09.	130-136	120-130	30	30000	8.33	0.02	1.10E-7	SiO_2	-	/	-	-
C5	08.10.	136	130	30	39300	10.92	0.9	2.20E-7	-	N,S	-	Al_2O_3 (a)	mono
C6	18.10.	136	130-138	-	39300	10.92	1.65	2.10E-7	SiO_2	2:1	mono	Al_2O_3 (a)	mono
C7	22.10.	136	138	30	39600	11	3.58	2.00E-7	SiO_2	-	mono	-	-
C8	30.10.	136	138	30	50100	13.92	5	1.80E-7	SiO_2	-	mono	-	-
C9	06.11.	136	138	100	86400	24	7.6	1.40E-7	SiO_2	S	-	Al_2O_3 (a)	ET
C10	12.11.	136	138	60	86400	24	7.98	1.30E-7	SiO_2	S	-	Al_2O_3 (a)	-
C11	16.11.	136	138	30	87600	24.33	7.48	1.20E-7	SiO_2	2:1	-	Al_2O_3	-

Table C.4: Co-evaporation of (ET)(TCNQ) and TCNQ

T_{e-t} denotes the evaporation temperature of (ET)(TCNQ), T_t the temperature of TCNQ and T_s the substrate temperature. Furthermore the deposition time t and the thickness d as displayed by the QMB are given. sub1 and sub2 denote the different substrates on which the samples where prepared with XRD1 and XRD2 referring to the corresponding Bragg scans on the respective substrate. The sulfur to nitrogen ratio S:N was determined with EDX measurements, which were carried out on the SiO_2 substrates, where possible. On Al_2O_3 charging effects inhibit meaningful results, as for sample C5, where sulfur and nitrogen could be detected, but the ratio could not be determined unambiguously. For sample C9 and C10 only sulfur was detected. The key for the result of the x-ray measurement (XRD1/2) is the following:
/ = no x-ray measurement performed, - = no peaks, mono = (ET)(TCNQ) monoclinic phase, ET = ET

Appendix D

Sputter parameters

D.1 Aluminum sputtering

For the sputtering of aluminum the turbo pump was slowed to 55% rotation speed, the UHV valve was opened 9 revolutions.

Oxygen flow parameters (Alicat Scientific mass flow controller):

- Mass flow = 5 sccm
- Volume flow = 55-60
- Temperature = $25-26$ °C
- Pressure = 0.85 PSIA

Sputter parameters (Hüttinger PFG 600 RF generator):

- CL = 54
- CT = 720
- DC = 600 V
- $P = 50$ W

Before sputtering, the aluminum target must be cleaned. Due to plasma oxidation or $SrTiO_3$ sputtering in the chamber, usually an insulating surface layer is formed on the aluminum target. To remove this layer often a higher sputter power is necessary. It is recommended to load a dummy sample to the chamber and clean the target before preparing an aluminum film.

D.2 Plasma oxidation

If the aluminum has been sputtered through a mask, it is recommended to unmount the mask before oxidation. This serves two purposes:

1. The mask will be protected from oxidation and corrosion.

2. The plasma reaches the thin film surface more easily and undisturbed by the metallic mask.

Oxygen flow parameters (Alicat Scientific mass flow controller):

- Mass flow = 15 sccm
- Volume flow = 53
- Temperature = $25 - 26$ °C
- Pressure = $4.2 - 4.4$ PSIA

Plasma parameters:

- $U = 470 - 500$ V
- $I = 10$ mA

Duration:

- $t = 2 - 3$ min

Pressure:

- $p = 0.380$ mbar

Bibliography

[1] J. Ferraris, D. Cowan, J. Walatka, and J. Perlstein, "Electron Transfer in a New Highly Conducting Donor-Acceptor Complex," *J. Am. Chem. Soc.*, vol. 95, no. 3, p. 948, 1973.

[2] C. K. Chiang, C. R. Fincher, J. W. Park, A. J. Heeger, H. Shirakawa, E. J. Louis S. C. Gau, and A. G. MacDiarmid, "Electrical Conductivity in Doped Polyacetylene," *Phys. Rev. Lett.*, vol. 39, no. 17, 1977.

[3] C. Dimitrakopoulos and D. Mascaro, "Organic thin-film transistors: A review of recent advances," *IBM J. of Research and Development*, vol. 45, no. 1, 2001.

[4] H. Mueller and Y. Ueba, "A novel, nonelectrochemical Synthesis of the organic superconductor κ-(BEDT-TTF)$_2$Cu(NCS)$_2$," *Bull. Chem. Soc. Jpn.*, vol. 66, no. 1, p. 32, 1993.

[5] N. Toyota, M. Lang, and J. Müller, *Low-Dimensional Molecular Metals*. Springer Series in Solid-State Science, first ed., 2007.

[6] M. Mizuno, A. Garito, and M. Cava, "'Organic Metals':Alkylthio Substitution Effects in Tetrathiafulvalene-Tetracyanoquinodimethane Charge-transfer Complexes," *J.C.S. Chem. Comm.*, 1978.

[7] T. Mori and H. Inokuchi, "Structural and electrical properties of (BEDT-TTF)(TCNQ)," *Sol. State Comm.*, vol. 59, no. 6, p. 355, 1986.

[8] H. Yamamoto, M. Hagiwara, and R. Kato, "New phase of (BEDT-TTF)(TCNQ)," *Synth. Met.*, vol. 133-134, p. 449, 2003.

[9] H. Urayama, H. Yamochi, G. Saito, K. Nozawa, T. Sugano, M. Kinoshita, S. Sato, K. Oshima, A. Kawamoto, and J. Tanaka, "A New Ambient Pressure Organic Superconductor Based on BEDT-TTF with T_c Higher than 10K," *Chem. Lett.*, p. 55, 1988.

[10] I. Giaever, "Energy Gap in Superconductors Measured by Electron Tunneling," *Phys. Rev. Lett.*, vol. 5, pp. 147–148, Aug 1960.

[11] I. Giaever, "Electron Tunneling Between Two Superconductors," *Phys. Rev. Lett.*, vol. 5, pp. 464–466, Nov 1960.

[12] A. D. McNaught and A. Wilkinson, *Compendium of Chemical Terminology: Iupac Recommendations: Gold Book*. IUPAC International Union of Pure and Applied Chem., second ed., 1997.

[13] R. Mulliken, "Molecular Compounds and their Spectra. II," *J. Am. Chem. Soc.*, 1952.

[14] P. F. Barbara, T. J. Meyer, and M. A. Ratner, "Contemporary Issues in Electron Transfer Research," *J. Phys. Chem.*, vol. 100, no. 31, p. 13148, 1996.

[15] R. A. Marcus, "Electron transfer reactions in chemistry. Theory and experiment," *Rev. Mod. Phys.*, vol. 65, pp. 599–610, Jul 1993.

[16] S. Gemming, M. Schreiber, and J.-B. Suck, *Materials for Tomorrow*. Springer, first ed., 2007.

[17] C. Strack, C. Akinci, V. Pashchenko, B. Wolf, E. Uhrig, W. Assmus, and M. Lang, " Resistivity studies under hydrostatic pressure on a low-resistance variant of the quasi-two-dimensional organic superconductor kappa-(BEDT-TTF)2Cu[N(CN)2]Br: Search for intrinsic scattering contributions," *Phys. Rev. B*, vol. 72, 2005.

[18] R. Zeis, *Single crystal field-effect transistors based on layered semiconductors*. PhD thesis, Universität Konstanz, 2005.

[19] W. D. Grobman and B. D. Silveman, "Intramolecular screening of crystal fields and the X-ray-photoemission determination of charge transfer in TTF-TCNQ," *Solid State Comm.*, vol. 19, p. 319, 1976.

[20] K. Kanoda, "Recent progress in NMR studies on organic conductors," *Hyperfine Interactions*, vol. 104, p. 235, 1997.

[21] J. Bednorz and K. Müller, "Possible High Tc Superconductivity in the Ba-La-Cu-O System," *Z. Phys. B*, vol. 64, p. 189, 1986.

[22] Y. F. Miura, S. Ohnishi, M. Hara, H. Sasabe, and W. Knoll, "Conductive thin films of bis(ethylenedithio)tetrathiafulvalene salt fabricated by a successive dry-wet process," *Appl. Phys. Lett.*, vol. 68, no. 17, p. 2447, 1996.

[23] S. Molas, J. Caro, J. Santiso, A. Figueras, J. Fraxedas, C. Meziere, M. Fourmigue, and P. Batail, "Thin molecular films of neutral tetrathiafulvalene-derivatives," *J. Chrystal Growth*, vol. 218, p. 399, 2000.

[24] A. Nollau, M. Pfeiffer, T. Fritz, and K. Leo, "Controlled n-type doping of a molecular organic semiconductor: Naphthalenetetracarboxylic dianhydride (NTCDA) doped with bis(ethylenedithio)-tetrathiafulvalene (BEDT-TTF)," *J. Appl. Phys.*, vol. 87, no. 9, p. 4340, 2000.

[25] H. Kobayashi, A. Kobayashi, Y. Sasaki, G. Saito, and H. Inokuchi, "The crystal and molecular structures of bis(ethylenedithio)tetrathiafulvalene," *Bull. Chem. Soc. Jpn*, 1986.

[26] T. Mori and H. Inokuchi, "Crystal of the Mixed-Stacked Salt of Bis(ethylenedithio)-tetrathiafulvalene (BEDT-TTF) and Tetracyanoquinodimethane (TCNQ)," *Bull. Chem. Soc. Jpn.*, 71987.

[27] J. Hunter, W. Massie, J. Meiklejohn, and J. Reid, "Thermal rearrangement in copper(II) thiocyanate," *Inorg. Nucl. Chem. Lett.*, vol. 5, p. 1, 1969.

[28] B. Ptaszynski, E. Skiba, and J. Krystek, "Thermal decomposition of Bi(III), Cd(II), Pb(II) and Cu(II) thiocyanates," *J. of Therm. Analysis and Calorimetry*, vol. 65, p. 231, 2001.

[29] H. Urayama, Y. H., G. Saito, S. Sato, A. Kawamoto J. Tanaka, T. Mori, Y. Maruyama, and H. Inokuchi, "Crystal Structure of Organic Superconductor, (BEDT-TTF)$_2$Cu(NCS)$_2$, at 298K and 104K," *Chem. Lett.*, p. 55, 1988.

[30] A. Schultz, M. Beno, G. U., H. Wang, A. Kini, and J. Williams, "Single-Crystal X-Ray and Neutron Diffraction Investigations of the Temperature Dependence of the Structure of the T_c = 10K Organic Superconductor κ-(BEDT-TTF)$_2$Cu(NCS)$_2$," *J. Sol. State Chem.*, vol. 94, p. 352, 1991.

[31] P. Guionneau, D. Chasseau, J. Howard, and P. Day, "Neutral bis(ethylenedithio)tetrathiafulvalene at 100 K," *Acta Chryst.*, vol. C56, 2000.

[32] K. Kanoda, "Electron Correlation, Metal-Insulator Transition and Organic Systems, (ET)$_2$X," *Physica C*, vol. 282-287, p. 299, 1997.

[33] K. Ichimura and K. Nomura, "d-wave Pair Symmetry in the Superconductivity of κ-(BEDT-TTF)$_2$X," *J. Phys. Soc. Jpn.*, vol. 75, no. 5, 2006.

[34] D. S. Acker, R. J. Harder, W. R. Hertler, W. Mahler, L. R. Melby, R. E. Benson, and W. E. Mochel, "7,7,8,8-tetracyanoquinodimethane and its electrically conducting anion-radical derivatives," *J. Am. Chem. Soc.*, vol. 82, no. 24, pp. 6408–6409, 1960.

[35] T. Hasegawa, K. Inukai, S. Kagoshima, T. Sugawara, T. Mochida, S. Sugiurac, and Y. Iwasa, "(BEDT-TTF)(F$_1$TCNQ) and (BEDT-TTF)(F$_2$TCNQ)$_x$(TCNQ)$_{1-x}$ (x ca. 0.5): all-organic metals down to 2 K," *Chem. Comm.*, p. 1377, 1997.

[36] T. Hasegawa, T. Mochida, R. Kondo, S. Kagoshima, Y. Iwasa, T. Akutagawa, T. Nakamura, and G. Saito, "(Mixed-stacked organic charge-transfer complexes with intercolumnar networks," *Phys. Rev. B*, vol. 62, no. 15, p. 10059, 2000.

[37] Y. Iwasa, K. Mizuhashi, and T. Koda, "Metal-insulator transition and antiferromagnetic order in bis(ethylenedithio)tetrathiafulvalene tetracyanoquinodimethane (BEDT-TTF)(TCNQ)," *Phys. Rev. B*, vol. 49, p. 3580, Feb. 1994.

[38] M. Miyashita, K. Uchiyama, H. Taniguchi, K. Satoh, Y. Uwatoko, N. Tajima, M. Tamura, and R. Kato, "High-pressure study of a Mott insulator (BEDT-TTF)(TCNQ)," *J. Phys. IV France*, vol. 114, p. 333, 2004.

[39] H. Yamamoto, N. Tajima, M. Hagiwara, R. Kato, and J.-I. Yamaura, "Strange Electric/Magnetic Behaviour of New (BEDT-TTF)(TCNQ)," *Synth. Met.*, vol. 135-136, p. 623, 2003.

[40] M. Kimata, Y. Oshima, K. Koyama, H. Ohta, M. Motokawa, H. Yamamoto, and R. Kato, "Fermi Surface Study of β -(BEDT-TTF)(TCNQ) by Magnetooptical Measurements," *Synth. Met.*, vol. 153, p. 369, 2005.

[41] H. Sakuma, M. Sakai, M. Iizuka, N. M., and K. Kudo, "Fabrication of Organic Transistors Using BEDT-TTF and (BEDT-TTF)TCNQ CT-Complex Films," *IEICE Trans. Electron.*, vol. E87, no. 12, p. 2049, 2004.

[42] M. Sakai, H. Sakuma, A. Saito, M. Nakamura, and K. Kudo, "Ambipolar field-effect transistor characteristics of (BEDT-TTF)(TCNQ) crystals and metal-like conduction induced by a gate electric field," *Phys. Rev. B*, vol. 76, p. 333, 2007.

[43] M. Karas, D. Bachmann, and F. Hillenkamp, "Influence of the Wavelength in High-Irradiance Ultraviolet Laser Desorption Mass Spectrometry of Organic Molecules," *Anal. Chem.*, vol. 57, p. 2935, 1985.

[44] M. Karas, D. Bachmann, U. Bahr, and F. Hillenkamp, "Matrix-assisted ultraviolet laser desorption of non-volatile compounds," *Int. J. of Mass Spectrometry and Ion Processes*, vol. 78, p. 53, 1987.

[45] F. Roth, *Organic field effect transistors*. PhD thesis, Universität Frankfurt am Main, 2009.

[46] P. Wagner, "Präparation epitaktischer Filme der Hochtemperatursupraleiter $YBa_2(Cu_{1-x}Zn_x)_3O_{7-\delta}$ und $Bi_2Sr_2CaCu2O_{8-\delta}$ mittels DC-Hochdruck-Sputtern und Messung charakteristischer Eigenschaften." Diploma Thesis (unpublished), Universität Darmstadt, 1991.

[47] M. Knudsen, "Die Gesetze der Molekularströmung und der inneren Reibungsströmung der Gase durch Röhren," *Annal. Phys.*, vol. 333, no. 1, p. 75, 1909.

[48] M. Knudsen, "Die Molekularströmung der Gase durch Öffnungen und die Effusion," *Annal. Phys.*, vol. 333, no. 5, p. 999, 1909.

[49] M. Knudsen, "Die Verdampfung von Kristalloberflächen," *Annal. Phys.*, vol. 357, no. 1, p. 105, 1917.

[50] M. Knudsen, "Das Cosinusgesetz in der kinetischen Gastheorie," *Annal. Phys.*, vol. 353, no. 24, p. 1113, 1916.

[51] D. P. Landau and K. Binder, *A Guide to Monte Carlo Simulations in Statistical Physics*. Cambridge University Press, 2000.

[52] M. Matsumoto and T. Nishimura, "Mersenne twister: a 623-dimensionally equidistributed uniform pseudo-random number generator," *ACM Trans. Model. Comput. Simul.*, vol. 8, no. 1, pp. 3–30, 1998.

[53] J. W. Ward, R. N. R. Mulford, and R. L. Bivins, "Study of Some of the Parameters Affecting Knudsen Effusion. II. A Monte Carlo Computer Analysis of Parameters Deduced from Experiment," *J. of Chem. Phys.*, vol. 47, no. 5, p. 1718, 1967.

[54] S. Adamson, C. O'Carroll, and J. F. McGilp, "The spatial distribution of flux produced by single capillary gas dosers," *Vacuum*, vol. 38, no. 4/5, p. 341, 1988.

[55] S. Adamson, C. O'Carroll, and J. F. McGilp, "The angular distribution of thermal molecular beams formed by single capillaries in the molecular flow regime," *Vacuum*, vol. 38, no. 6, p. 463, 1988.

[56] S. Adamson and J. F. McGilp, "Monte carlo calculations of the beam flux distribution from molecular-beam epitaxy sources," *J. Vac. Sci. Technol. B*, vol. 7, no. 3, p. 487, 1989.

[57] J. W. Ward and M. V. Fraser, "Study of Some of the Parameters Affecting Knudsen Effusion. VI. Monte Carlo Analyses of Channel Orifices," *J. of Chem. Phys.*, vol. 50, no. 4, p. 1877, 1969.

[58] R. E. Stickney, R. R. Keating, S. Yamamoto, and W. J. Hastings, "Angular distribution of flow from orifices and tubes at high knudsen numbers," *J. Vac. Sci. Technol.*, vol. 4, no. 1, p. 10, 1966.

[59] Z. R. Wasilewski, G. Aers, A. J. SpringThorpe, and C. J. Miner, "Growth uniformity studies in molecular beam epitaxy," *J. of Crystal Growth*, vol. 111, p. 70, 1991.

[60] W. Gericke, M. Höricke, and J. von Kalben, "A detailed study of the molecular beam flux distribution of MBE effusion sources," *Vacuum*, vol. 42, no. 18, p. 1209, 1991.

[61] S. Kirsch, J. Griesche, and W. Gericke, "Monte-Carlo Simulation of the Beam Flux Distribution of Molecular-Beam Epitaxy Sources by Consideration of Intermolecular Collisions," *Phys. stat. Sol.*, vol. 123, p. 441, 1991.

[62] M. Atterer, R. Meyer, F. Peters, L. Gmelin, E. Pietsch, and M. Becke-Goehring, *Gmelins Handbuch der anorganischen Chemie, System-Nr. 60, Kupfer Teil B, Lfg. 2*. Verl. Chemie, eighth ed., 1961.

[63] J. Fraxedas, S. Molas, A. Figueras, I. Jiménez, R. Gago, P. Auban-Senzier, and M. Goffman, "Thin Films of Molecular Metals:TTF-TCNQ," *Journal of Solid State Chemistry*, vol. 168, p. 384, 2002.

[64] C. Rojas, J. Caro, M. Grioni, and J. Fraxedas, "Surface characterization of metallic molecular organic thin films: tetrathiafulvalene tetracyanoquinodimethane," *Surface Science*, vol. 482-485, p. 546, 2001.

[65] C. Grimm, "Electronic transport in nanocomposites made by electron beam induced deposition." Diploma Thesis (unpublished), Universität Frankfurt, 2006.

[66] D. Klingenberger, "Supraleitung in wolframhaltigen Nanokompositstrukturen." Diploma Thesis (unpublished), Universität Frankfurt, 2008.

[67] L. Esaki, "Long journey into tunneling," *Nobel Lecture*, 1973.

[68] L. Esaki and P. J. P. J. Stiles, "Study of electronic band structures by tunneling spectroscopy: Bismuth," *Phys. Rev. Lett.*, vol. 14, no. 22, p. 902, 1965.

[69] L. Esaki and P. J. P. J. Stiles, "New type of negative resistance in barrier tunneling," *Phys. Rev. Lett.*, vol. 16, no. 24, p. 1108, 1966.

[70] G. Blendin, "Structural and transport properties of $CeCoIn_5$ thin films." Diploma Thesis (unpublished), Universität Frankfurt, 2006.

[71] J. Simmons and G. Unterkofler, "Potential Barrier Shape Determination in Tunnel Junctions," *J. Appl. Phys.*, vol. 34, p. 1828, 1963.

[72] I. Utke, V. Friedli, S. Amorosi, J. Michler, and P. Hoffmann, "Measurement and simulation of impinging precursor molecule distribution in focused particle beam deposition/etch systems," *Microelectronic Engineering*, vol. 83, p. 1499, 2006.

Thanks!

Als erstes möchte ich Prof. Dr. Michael Huth danken, für die Möglichkeit diese Arbeit in seiner Gruppe durchzuführen. Gerade weil es zu Anfang eine kleine "neue" Arbeitsgruppe war, hat sich mit dem Laboraufbau auch ein besonderer Gruppenzusammenhalt aufgebaut.

Michael Huth persönlich möchte ich für seine kritischen Diskussionen, aber auch für seine Offenheit gegenüber allen Vorschlägen danken. Von seinem unglaublichen Wissen kann man jede Menge lernen und trotzdem (oder gerade deswegen?) kann man selbst die dümmsten Fragen stellen ohne sich zu blamieren.

Natürlich gilt der Dank auch der gesamten Arbeitsgruppe:
Danke an Gabi und Martin für die nette Bürogemeinschaft.
Danke an Tini, Dirk und Jörg, den Hüter der Kaffeemaschine, für die Bereitstellung des inoffiziellen Gruppentreffpunktes!
Vielen Dank an Tini, Dirk und Martin für die strukturierten Siliziumproben - und an Tini ein besonderer Dank für die REM Bilder und EDX Messungen!
Danke auch an Fabrizio, Oleksiy, Oleksander und Harald für eine gute Zeit.

Vielen Dank an die elektronische Werkstatt für ihre unermüdliche Fehlersuche bei unserem alten Massenspektrometer und für Kabelnachschub jeglicher Art.

Ganz besonders möchte ich mich auch bei der mechanischen Werkstatt bedanken, ohne deren unermüdlichen Einsatz mindestens die Hälfte dieser Arbeit gar nicht möglich gewesen wäre. Vor allem Herr Pfeiffer und Herr Hohmann sollen erwähnt werden, die stets auch noch für die absurdeste Idee eine Lösung finden und auch in der Mittagspause mal ein Snowboard reparieren!

Ich bedanke mich auch bei allen Mitdoktoranden, ganz nach dem Motto: Geteiltes Leid ist halbes Leid!
Ebenfalls bedanke ich mich beim gesamten Physikalischen Institut für eine gute Zusammenarbeit und dafür, dass bei Problemen immer jemand ein offenes Ohr hat.

Danke auch an Jan C. Bernauer für seine 24 h-TEX-Beratung.

Außerdem möchte ich mich bei meiner Familie bedanken - bei meiner Schwester, dass sie den Kampf mit dem Rechtschreibteufel aufgenommen hat und für ihre kostenlose "chemische" Beratung. Bei meinen Eltern bedanke ich mich für Ihre uneingeschränkte Unterstützung!

Als letztes möchte ich mich bei Florian bedanken für all das, was wir während dieser Arbeit zusammen erlebt haben und für seine Aufmunterungen, wenn ich nicht mehr weiter wusste - tack min lilla vän!

Die VDM Verlagsservicegesellschaft sucht für wissenschaftliche Verlage abgeschlossene und herausragende

Dissertationen, Habilitationen, Diplomarbeiten, Master Theses, Magisterarbeiten usw.

für die kostenlose Publikation als Fachbuch.

Sie verfügen über eine Arbeit, die hohen inhaltlichen und formalen Ansprüchen genügt, und haben Interesse an einer honorarvergüteten Publikation?

Dann senden Sie bitte erste Informationen über sich und Ihre Arbeit per Email an *info@vdm-vsg.de*.

Sie erhalten kurzfristig unser Feedback!

VDM Verlagsservicegesellschaft mbH
Dudweiler Landstr. 99
D - 66123 Saarbrücken

Telefon +49 681 3720 174
Fax   +49 681 3720 1749

www.vdm-vsg.de

Die VDM Verlagsservicegesellschaft mbH vertritt

Printed by Books on Demand GmbH, Norderstedt / Germany